AF598141

Bioplastics within the Circular Bioeconomy

Edited by Mohamed Samer

Published in London, United Kingdom

Bioplastics within the Circular Bioeconomy
http://dx.doi.org/10.5772/intechopen.1003551
Edited by Mohamed Samer

Contributors
Elizabeta Hernández Domínguez, Emmanuel Flores Huicochea, Fatma Abdelrahman, Hassan Abdellatif, Heba Younis, Mariam Amer, María de la Luz Sánchez Mundo, Mohamed Said Mahmoud, Mohamed Samer, Rosalía América González Soto, Sungeun Ahn

First published in London, United Kingdom, 2025 by IntechOpen
IntechOpen is the global imprint of INTECHOPEN LIMITED, registered in England and Wales, registration number: 11086078, 167-169 Great Portland Street, London, W1W 5PF, United Kingdom

For EU product safety concerns: IN TECH d.o.o., Prolaz Marije Krucifikse Kozulić 3, 51000 Rijeka, Croatia, info@intechopen.com or visit our website at intechopen.com.

British Library Cataloguing-in-Publication Data
A catalogue record for this book is available from the British Library

Bioplastics within the Circular Bioeconomy
Edited by Mohamed Samer
p. cm.
Print ISBN 978-0-85014-951-7
Online ISBN 978-0-85014-950-0
eBook (PDF) ISBN 978-0-85014-952-4

If disposing of this product, please recycle the paper responsibly.

Meet the editor

Prof. Dr. Mohamed Samer is a Laureate of the Cairo University Scientific Excellence Award (2019). He is the recipient of The Privilege Medal of First Class (2017), issued by the Egyptian President. He is also a Laureate of the State Encouragement Award (2016) conferred by the Egyptian Government. He is listed in the Stanford/Elsevier World's Top 2% Scientists List (2023 and 2024). He was granted his Ph.D. by the University of Hohenheim in Germany. He has led 19 research projects and established five laboratories. With 140 publications to his credit, he is an active member of 13 scientific societies. He also serves as a peer reviewer for 30 high-impact journals, having reviewed 130 papers, and as a reviewer for eight funding agencies, where he has evaluated 40 research project proposals.

Contents

Preface

This book aims to provide useful information on bioplastic production from biomass, i.e., biological wastes, known as biowastes, generated from agriculture, forestry, and the food industry, which are largely untapped sources for producing value-added bioproducts such as bioplastics.

Their recovery utilizes biological and chemical processes that provide alternative sources for feedstock in the production of different types of bioplastics. For instance, biopolymers can recover from biomass using biorefineries. Similarly, many opportunities exist for petroplastic alternatives that can be produced from biomass using biorefineries. Resource biorecovery thus supports sustainability goals by reinjecting bioproducts, especially bioplastics, into the circular bioeconomy. Authors have contributed up-to-date knowledge from basics to apex, allowing readers to comprehend the topic more deeply.

This book is intended to be a helpful resource for NGOs, universities, research institutes and centers, experimentalists, academics, scientists, scholars, and researchers, as well as undergraduate and graduate students worldwide who specialize in energy engineering, environmental engineering, bioresource engineering, biosystems engineering, microbiology, biochemistry, and biotechnology.

Dr. Mohamed Samer
Professor,
Faculty of Agriculture,
Department of Agricultural Engineering,
Cairo University,
Giza, Egypt

Chapter 1

Introductory Chapter: Bioplastics as Substitutes to Petroplastics

Mohamed Samer and Mariam Amer

1.Introduction

Plastics are utilized practically everywhere, including in everyday household packaging materials, bottles, printers, cell phones, and more. Additionally, it is used by manufacturing sectors including pharmaceuticals and autos. Since single-use plastics contribute approximately 60% of the yearly production of plastic garbage, there have been focused efforts in recent years to reduce their use [1, 2]. Although plastics have many benefits, they also have some drawbacks. For example, because they do not break down quickly, the original products may remain in the environment for hundreds or even thousands of years, leading to pollution and a significant environmental problem. In addition, the majority of plastics come from fossil fuels, which are finite and non-renewable. These natural resources are running out extremely quickly due to the overproduction of plastic and its waste products [1, 3]. Owing to their large greenhouse gas emissions and high carbon footprint, petroplastics produced from the petrochemical industries are not sustainable anymore [4]. For this reason, there is growing interest in identifying materials with similar qualities to petroleum-based plastics to manage plastic waste on Earth by identifying environmentally friendly substitutes [1, 2].

2. Why bioplastics?

Bioplastics are substitutes to the regular petroplastics [5]. The negative environmental impacts of producing as well as recycling petroplastics can be avoided through the conversion of biomass into bioplastics [6]. Additionally, bioplastics have enormous number of uses and applications, besides a large economic opportunity [7]. A plastic material is considered a bioplastic if it is either bio-based, biodegradable, or possesses both qualities, according to European Bioplastics [8]. Bioplastics are a family of materials with a variety of uses and characteristics. They are recyclable but may or may not be biodegradable. The mechanical characteristics are fairly comparable to those obtained from fossils. For instance, bio-sourced PE and bio-PVC from sugarcane [1, 2]. One of the most sustainable carbon upcycling viewpoints that is gaining a lot of attention these days is the utilization of greenhouse gases, such as carbon dioxide, for the production of bioplastic. In the long run, it will encourage the creation of bioplastics with a low carbon impact [2].

3. What are bioplastics?

Bioplastics are biopolymers processed from biomass and are biodegradable [5]. The development of bioplastics now relies heavily on agricultural products for its raw materials, which indirectly threatens food security [9]. Therefore, utilizing biologically derived organic wastes will help reduce our reliance on crops and could perhaps help manage waste more efficiently [9].

Bioplastics can be produced from biomass such as agriculture sector residues and food waste through bioconversion processes. For instance, bioplastics can be produced from potato peels and banana peels [5, 7] as well as wastes rich in starch and carbohydrates (**Figure 1**).

There are several types of bioplastics (**Figure 2**) such as polyhydroxyalkanoates (PHA), polylactic acid (PLA), and polybutylene succinate (PBS), which are bio-based

Figure 1.
A photograph showing a piece of raw bioplastic produced from potato peels.

Overview of Bioplastics

Proteins
Biodegradable bioplastics from plant-based proteins

Polyhydroxyalkanoates
Produced through bacterial fermentation of sugars

Cellulose
Extracted from cotton and wood, used in specialized applications

Polyhydroxybutyrate
Used in various consumer products, safe for human use

Starch
Derived from common crops, used in packaging

Polylactic acid
Known for being biodegradable and thermally stable

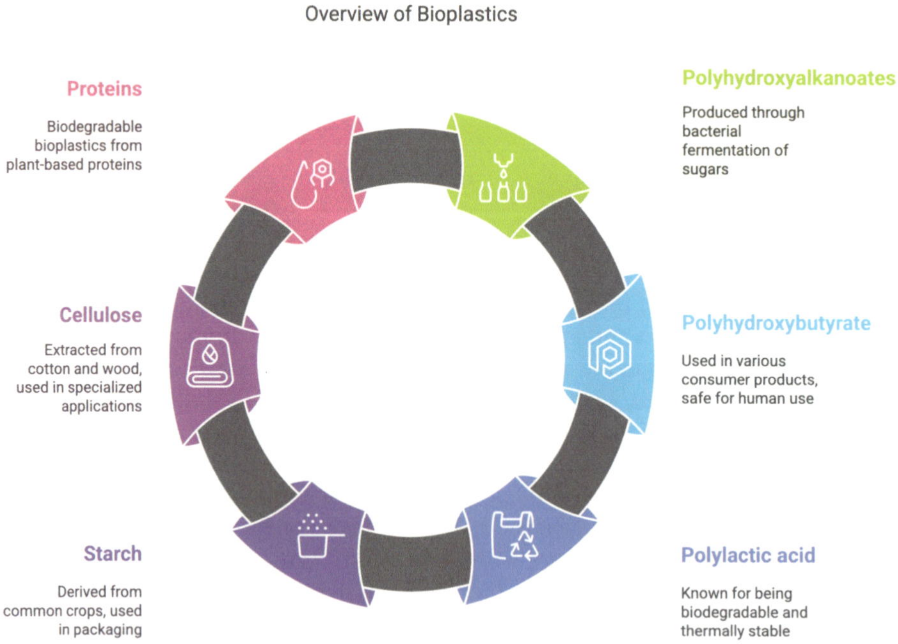

Figure 2.
Types of bioplastics.

and compostable [10]. Thermoplastic starch (TPS), polylactides (PLA), poly-β-hydroxybutyric acid (PHB), and its co-polymers (PHAs) are the supreme substantial biopolymers presently appealing interest.

As an exceptional combination of life sciences and engineering, the multidisciplinary research field of synthetic biology can provide novel pathways for the reformation of biosynthesis methods for biomass processing harmoniously and can ultimately develop low-cost, effective bioprocesses of biomass into decomposable bioplastics [11].

4. What is circular bioeconomy?

A circular bioeconomy is a nature-driven bio-based economy. It is a new-fangled economic model that points up the usage of renewable bioresources and concentrates on minimalizing waste and substituting the multitude of varieties of fossil-based non-renewable products presently in consumption.

The circular bioeconomy is the nexus of the two new ideas—the bioeconomy and the circular economy. **Figure 3** summarizes the discussion surrounding the connection between the bioeconomy and the circular economy [12].

The bioeconomy provides a comprehensive, multisectoral strategy that has a great chance of mitigating climate change in different ways. There are numerous opportunities to lower anthropogenic greenhouse gas (GHG) emissions, including storing carbon in bio-based products, storing carbon dioxide (CO_2) from the atmosphere in

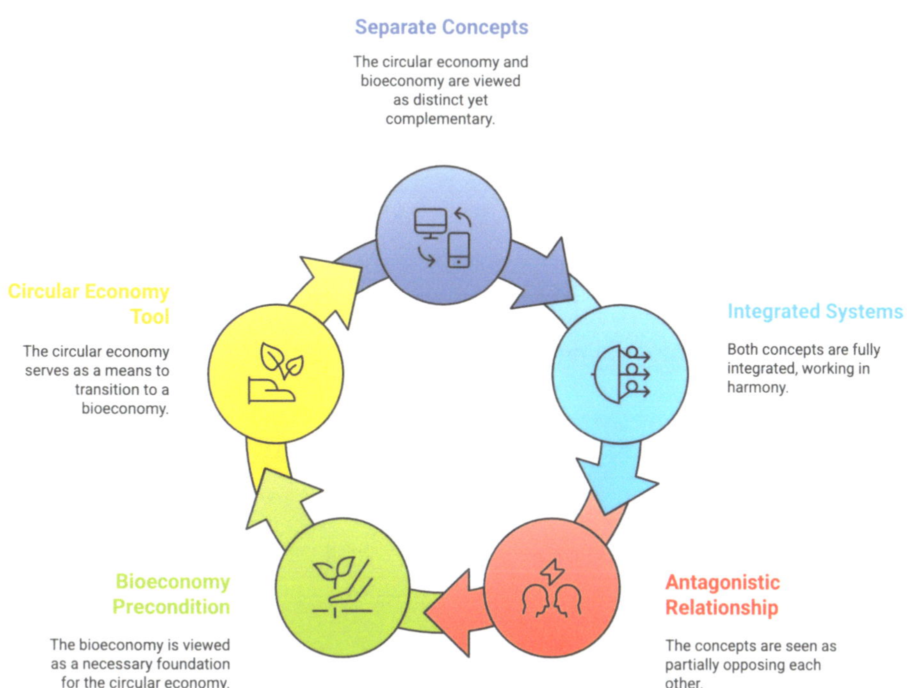

Figure 3
Relation between the bioeconomy and the circular economy.

biomass from plants and microorganisms, and replacing fossil fuel-based feedstocks, inputs, and products with bio-based ones [13, 14].

Petroplastics are one of these products that ought to be replaced by eco-friendly bioplastics, which can convert up to 90% of the organic carbon into CO_2 in 180 days. After the product's useful life, the carbon can return to the biosphere as CO_2 [12].

5. Conclusion

Humans, wildlife, marine life, and the environment are all at risk from synthetic plastics derived from petrochemicals, as is the quick buildup of plastic garbage. Considering the extent to which the production of synthetic plastics will eventually deplete petrochemical resources and pollute the environment worldwide.

Utilizing organic waste from biological origins for bioplastic production not only lessens our reliance on edible feedstock but also effectively aids in solid waste management as part of an expanding circular bioeconomy. Biodegradable bioplastics can have the same qualities as traditional plastics while also offering additional benefits due to their low carbon footprint (e.g., full biodegradability to CO_2 without harmful byproducts).

Author details

Mohamed Samer[1*] and Mariam Amer[2]

1 Department of Agricultural Engineering, Faculty of Agriculture, Cairo University, Egypt

2 Agricultural Engineering Research Institute (AEnRI), Agricultural Researcher Center (ARC), Egypt

*Address all correspondence to: msamer@agr.cu.edu.eg

References

[1] Shamsuddin IM, Jafar AJ, Shawai ASA, Yusuf S, Lateefah M, Aminu I. Bioplastics as better alternative to petroplastics and their role in national sustainability: A review. Advances in Bioscience and Bioengineering. 2017;**5**(4):63-70. DOI: 10.11648/j.abb.20170504.13

[2] Saharan R, kharb, J. Exploration of bioplastics: A review. Oriental Journal of Chemistry. 2022;**38**(4):840-854. DOI: 10.13005/ojc/380403

[3] Falzarano M, Marìn A, Cabedo L, Polettini A, Pomi R, Rossi A, et al. Alternative end-of-life options for disposable bioplastic products: Degradation and ecotoxicity assessment in compost and soil. Chemosphere. 2024;**362**:142648. DOI: 10.1016/j.chemosphere.2024.142648

[4] Semba T, Sakai Y, Sakanishi T, Inaba A. Greenhouse gas emissions of 100% bio-derived polyethylene terephthalate on its life cycle compared with petroleum-derived polyethylene terephthalate. Journal of Cleaner Production. 2018;**195**:932-938. DOI: 10.1016/j.jclepro.2018.05.069

[5] Samer M, Hijazi O, Mohamed BA, Abdelsalam E, Amer M, Yacoub IH, et al. Environmental impact assessment of bioplastics production from agricultural crop residues. Clean Technologies and Environmental Policy. 2022;**24**(3):815-827

[6] Samer M, Khalef ZF, Abdelall TM, Moawya WM, Farouk AH, Abdelaziz SA, et al. Bioplastics production from agricultural crop residues. Agricultural Engineering International: CIGR Journal. 2019;**21**(3):190-194

[7] Samer M. Bioplastics Production from Agricultural Wastes. ISBN 978-613-9-93557-4,. Germany: LAP LAMBERT Academic Publishing; 2018

[8] Nizamuddin S, Baloch AJ, Chen C, Arif M, Mubarak NM. Bio-based plastics, biodegradable plastics, and compostable plastics: Biodegradation mechanism, biodegradability standards and environmental stratagem. International Biodeterioration & Biodegradation. 2024;**195**:105887. DOI: 10.1016/j.ibiod.2024.105887

[9] George N, Debroy A, Bhat S, Singh S, Bindal S. Biowaste to bioplastics: An ecofriendly approach for a sustainable future. Journal Of Applied Biotechnology Reports. 2021;**8**(3):221-233. DOI: 10.30491/JABR.2021.259403.1318

[10] Ali Z, Abdullah M, Yasin MT, Amanat K, Ahmad K, Ahmed I, et al. Organic waste-to-bioplastics: Conversion with eco-friendly technologies and approaches for sustainable environment. Environmental Research. 2024;**244**:117949. DOI: 10.1016/j.envres.2023.117949

[11] Tsang YF, Kumar V, Samadar P, Yang Y, Lee J, Ok YS, et al. Production of bioplastic through food waste valorization. Environment International. 2019;**127**:625-644. DOI: 10.1016/j.envint.2019.03.076

[12] Tan EC, Lamers P. Circular bioeconomy concepts—A perspective. Frontiers in Sustainability. 2021;**2**:701509. DOI: 10.3389/frsus.2021.701509

[13] Karan H, Funk C, Grabert M, Oey M, Hankamer B. Green bioplastics as part of a circular bioeconomy. Trends in Plant Science. 2019;**24**(3):237-249. DOI: 10.1016/j.tplants.2018.11.010

[14] Gomez San Juan M, Harnett S, Albinelli I. Sustainable and Circular Bioeconomy in the Climate Agenda: Opportunities to Transform Agrifood Systems. Rome, FAO; 2022. DOI: 10.4060/cc2668en

Chapter 2

Agro-Industrial Alchemy: Transforming Waste into Wealth with Bio-Based Plastics

Elizabeta Hernández Domínguez,
María de la Luz Sánchez Mundo,
Rosalía América González Soto
and Emmanuel Flores Huicochea

Abstract

The innovative conversion of agro-industrial wastes, such as sugarcane bagasse, corn husks, and fruit peels, into valuable bioplastic materials is explored, contributing to sustainable industrial practices. The focus is on utilizing residues from agricultural and industrial processes, typically underutilized or discarded, as raw materials for producing environmentally friendly plastics. The scope includes examining the types of agro-industrial wastes suitable for bioplastic production, the technological advancements enabling this transformation, and the environmental and economic impacts. Specifically, it discusses how these bioplastics can significantly reduce greenhouse gas emissions compared to traditional plastics and their potential cost-effectiveness in the long term. It also addresses the challenges and opportunities in scaling these practices, the role of policy in supporting waste-to-wealth initiatives, and the potential of these bioplastics to integrate into and enhance the circular economy. This chapter aims to provide a comprehensive understanding to academics, industry professionals, and policymakers on how bioplastics from agro-industrial waste can pave the way for more sustainable manufacturing processes.

Keywords: bioplastics, agro-industrial waste, sustainability, waste valorization, circular economy

1. Introduction

The middle of the XIX century began the plastic period. Alexander Parkes synthesized a semisynthetic plastic cellulose nitrate named Parkesine, and then, the celluloid was produced by Wesly Hyatt [1]. The main characteristics of synthetic polymers are high molecular weight, lightweight, durability, flexibility, durability, and so on [2, 3]. Today, these properties have made plastics widely used in art, houseware, biomedicine, industry, and packaging [4–8].

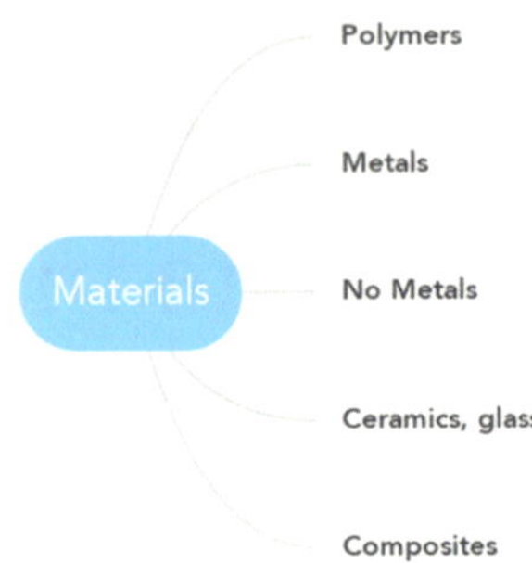

Figure 1.
Classification of material according to Science Material. The authors' chapter elaborated the figure.

In general, there are five types of substances in material science (**Figure 1**). Plastic is the material that has the most significant production of all kinds of material. The global production of polymers has increased exponentially since 1950, and 300 million tons are produced annually [9]. There is an estimation that between 9 and10% of plastic waste is recycled, 12% is incinerated, and 79% of waste plastic accumulates in landfills or the natural environment. Between 4.8 and 12.7 million tons of plastics enter the ocean yearly as macroscopic debris and microplastic particles (**Figure 2**) [10–13].

In principle, the main characteristics of synthetic polymers (lightweight, chemical resistant, and cheap) that increase their use in many areas of human life have implied the contamination of ecosystems because of low rate of degradation or just only fragmentation. Plastic contamination impacts the marine ecosystem; it is estimated that millions of tons annually enter the oceans. Many aquatic animals (fishes, seabirds, and mammalians) will die from ingesting plastic fragments. The plastic debris absorbs chemical contaminants from the water, and then, the aquatic animals ingest them, and the compound toxics are incorporated into the trophic chain [14–17].

A bioplastic refers to a polymer derived from renewable resources with the unique capability of decomposing by microorganisms with the aid of enzymatic processes, resulting in carbon dioxide, water, organic substances, and inorganic compounds. The categorization of biodegradable polymers primarily revolves around two distinct classes: agro-polymers, which encompass starch, proteins, and chitin, and biopolyesters, consisting of polyhydroxyalkanoates and polylactic acid. These classifications play a crucial

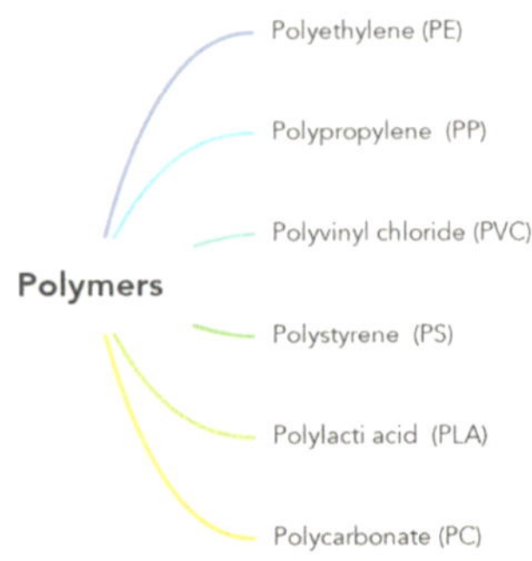

Figure 2.
Main polymers produced worldwide. The authors' chapter elaborated the figure.

role in understanding the diverse nature of bioplastics and their potential applications in various fields, such as environmental sustainability, packaging, and biomedical industries [18, 19]. Bioplastics is a promising alternative to conventional petroleum-based plastic. The first advantage is replacing the traditional plastic and reducing the hazardous components released to the environment due to the fragmentation or degradation of oil-based plastic because they are designed to break down in the environment [20].

Bioplastic research began in the XX century. In 1926, the scientific French Maurice Lemoigne found the first bioplastic poli(3-hidroxibutirato), called PHB, produced from bacteria, but PHB was ignored because of its low cost and abundance of oil. The second resurgence was during the 1970s in the last century during the oil crisis [13, 21]. The social concert about plastic contamination and its effects opened commercial opportunities and in 1990 began the commercial production of bioplastic [21]. In 1997, Cargill and Dow Chemicals formed NatureWorks, which produces polylactic acid (PLA) [22, 23], and today, it is the primary producer of PLA.

At first, bioplastic production used PLA as the main component. Subsequently, other elements, such as polysaccharides (starch) and proteins (whey), were incorporated to produce bioplastics with improved properties, but polyhydroxyalkanoates(PHAs) and polyhydroxy butyrates PHBs obtained from bacteria are also used [24]. Several research papers have appeared using agro-industrial waste without or with mechanical or chemical modifications to blend with the traditional components to produce bioplastics.

Agro-industrial waste, commonly called agro-waste, refers to the residual materials produced during the cultivation and processing of unrefined agricultural commodities such as fruits, vegetables, meat, poultry, dairy products, and crops. These discarded materials include a variety of substances such as animal excrement, agricultural remnants (for instance, corn stalks and sugarcane bagasse), leftovers from food processing, as well as harmful agents like pesticides and herbicides, which are secondary to the main outputs of farming operations and often exhibit reduced economic worth in comparison to the expenses associated with their gathering, transportation, and preparation for utilization.

The management and appropriate disposal of agro-waste are crucial aspects that necessitate attention due to the environmental impacts and potential health hazards of mishandling such materials [25, 26]. In this scenario, agro-industrial waste pertains to the waste generated from residual materials from unprocessed agricultural commodities, highlighting the importance of efficiently managing and utilizing these byproducts to minimize environmental impact and promote sustainability in the farming sector.

The agricultural sector involves a wide range of byproducts produced while cultivating and processing different cultivation commodities. These wastes include crop residues such as wheat straw, paddy straw, corn stalks, sugarcane bagasse, corn cob, and rice husk. These residues are rich in polysaccharides such as cellulose, hemicellulose, and lignin; this mixture is usually known as lignocellulose material [27–29].

The chapter's main objective is to analyze the principal sources of agricultural waste worldwide, examine how these residues produce today's bioplastics, and show commercial products.

2. Agro-industrial waste as a resource

At a global level, the main crops belong to the grass family, scientifically named *Poaceae.* This family is dominant in modern agriculture and includes over 10,000 species, many of which are crucial for global food security and economies. Key cereal crops such as rice, wheat, maize, sorghum, barley, oats, and millet are all members of the *Poaceae* family. These crops are vital for human consumption and play a significant role in animal feed and other agricultural applications. The economic importance of the *Poaceae* family is underscored by the fact that maize, wheat, and rice alone account for a significant portion of the world's food supply [30].

Figure 3 shows the highest global crop production over a long period, and the first four are crops that belong to the *Poaceae* family. From now on, we will focus the analysis only on these agricultural wastes and mainly on lignocellulosic waste.

The processing of sugarcane produces bagasse and other subproducts; almost 15 to 25% of the total weight of sugarcane corresponds to bagasse [32, 33]. Due to the enormous amount of bagasse used during sugarcane processing today, the traditional uses are fuel and animal food. Alternatively, it can produce bioethanol, activated carbon, biodegradable material, and packaging products [33–37]. In the construction sector, bagasse can be used as an aggregate to make building blocks. Composites made with bagasse and binders show mechanical strength for non-load-bearing walls and offer better thermal insulation compared to traditional materials [38].

In the case of corn, during the harvest, it produces around 2 ton of waste per 1 t of corn harvest; the waste includes stalks, roots, leaves, and corn cobs [39]. About 30 to 33% of waste is generated per ton of rice harvest, primarily rice straw and rice husk, the last being the most abundant [40, 41]. For the remaining crop, for each ton of wheat harvest, a ton of waste is produced [42–44].

Figure 4 shows the estimated amount produced annually for the main crops. Despite sugarcane being the most produced, it is the third primary waste producer. The critical waste producer is corn. Despite the traditional practice of leaving corn

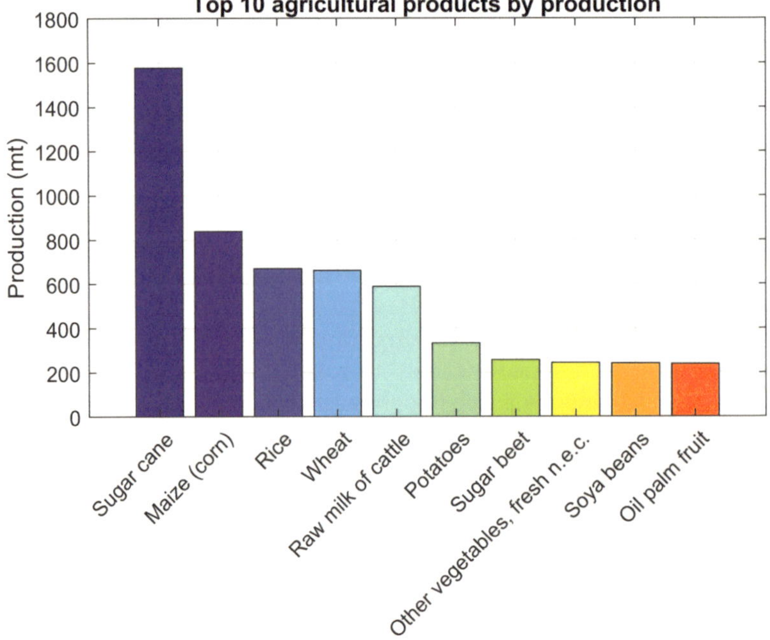

Figure 3.
Average annual production of the main crops globally from 1994 to 2022. Production is reported in millions of tons [31]. The authors' chapter elaborated the figure.

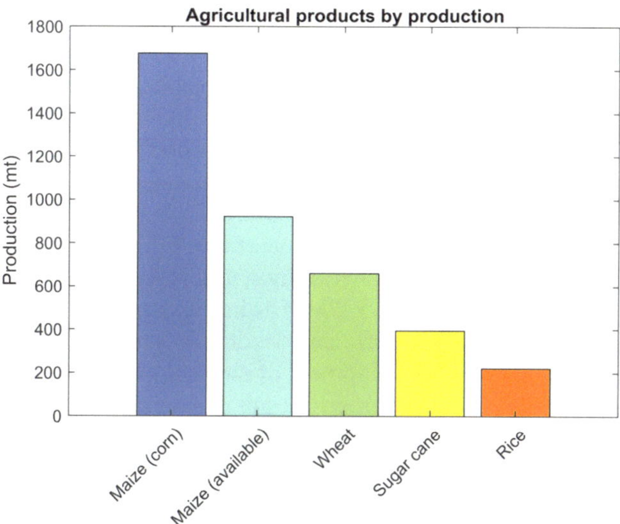

Figure 4.
Average annual agricultural waste production for the most produced crops globally for 1994–2022. Production is reported in millions of tons. The waste amount was estimated using several sources [31–33, 39–44]. The amount of corn waste available was calculated after considering the percent of waste burning in the field. The authors' chapter elaborated the figure.

waste in the field and the global practice of burning crop residues, the available waste residues are about 55%, corresponding to around 922 million tons annually (see **Figure 4**) [45, 46]. The potential use of the three principal waste producers is around 1976 million tons.

Biorefineries represent a transformative approach to biomass utilization, integrating various technologies to convert raw materials into diverse products, including fuels, chemicals, and materials. Central to this concept is the use of lignocellulosic feedstock, which allows for the sustainable production of multiple outputs from renewable resources, thereby enhancing economic viability and resource efficiency [47], and the circular economy is an innovative financial model that seeks to harmonize economic activity with environmental sustainability by redesigning processes and promoting the cycling of materials. This approach aims to minimize waste and maximize resource efficiency, thereby contributing to sustainable business models that prioritize ecosystem functioning and human well-being [48, 49].

Developing concepts like biorefinery and circular economy play a crucial role in converting agro-industrial waste into value-added chemicals such as furan derivatives, and levulinic acid, and these building blocks are essential for industrial applications.

Now, we focus on analyzing agricultural waste's use to convert it into bioplastics, mainly corn, wheat, and sugarcane bagasse waste, see **Figure 4**. Agricultural wastes, typically, can be transformed into bioplastic because they contain lignocellulosic polymers like lignin, hemicellulose, and cellulose. These natural polymers break down with the help of microbes, which makes them eco-friendly options compared to traditional plastics. Furthermore, their abundance and biodegradability should be considered [50, 51]. From this economic point of view, the commercial manufacturing capacity of bioplastics is projected to reach 7.6 million tons by 2026, with Europe holding a substantial 43.5%. The demand for bioplastics is growing at an annual rate of 10–20%, partly due to the scarcity and rising prices of oil [18, 52].

3. Technological advancements in bioplastic production

The production of bioplastics has seen significant technological advances, driven by the urgent need to address environmental pollution caused by traditional plastics. One notable development is using renewable bio-based feedstocks such as plant polysaccharides, starch, vegetable oils, and lignocellulosic biomass, which are abundant and cost-effective sources for bioplastic synthesis [53]. Advances in microbial technology, remarkably through metabolic engineering, have enabled the efficient production of bioplastics by utilizing microorganisms to naturally synthesize plastics or their monomeric building blocks from biomass substrates [53]. The incorporation of various organic acids and plasticizers, such as citric acid, acetic acid, lactic acid, glycerin, polyvinyl alcohol, and nanomaterials, has led to the development of durable and economical biopolymers with enhanced physico-mechanical properties, suitable for mass production [54].

Additionally, the use of microbial fermentation to produce polyhydroxyalkanoates (PHAs) has resulted in biopolymers with characteristics comparable to synthetic plastics, further expanding their commercial applications [52]. Another promising approach is the semi-biosynthetic route for producing poly(ethylene furanoate) (PEF) monomers *via* microbial-based biorefinery and subsequent catalyst-mediated polycondensation. However, it faces challenges such as high pretreatment costs and the need for efficient substrate utilization and pathway optimization [55].

The development of bioplastics has also benefited from advancements in strain development, genome sequencing, and editing technologies, which have accelerated research efforts and improved the production and biodegradation of bioplastics [56]. Various production methods, including polymerization, injection molding, film casting, 3D printing, solvent-free methods, and electrospinning, have been explored to enhance the properties and applications of bioplastics, particularly in medical equipment, food packaging, textiles, and healthcare appliances [57]. Including different fillers has been shown to improve bioplastics' mechanical and barrier properties, making them more suitable for food packaging applications [58].

Furthermore, the discovery and engineering of biological catalysts capable of depolymerizing a wide range of plastics have opened new avenues for plastic recycling and upcycling, with significant improvements in enzyme stability and functionality through protein engineering and immobilization techniques [59, 60]. These technological advances collectively contribute to the growing market share of bioplastics, projected to reach 7.6 million tons by 2026, with Europe currently holding a significant market share [52]. The continued exploration of cleaner production methods, end-of-life management, and toxicity assessment of degradation products is crucial to ensuring bioplastics' economic viability and environmental sustainability as alternatives to synthetic plastics [52].

One notable breakthrough is the development of biodegradable plastics with superior mechanical performance by incorporating lignin–Cu nanoparticles (LNP@Cu) into poly(vinyl alcohol) (PVA), mimicking the granular nanostructure in mussel byssus cuticle. This innovation resulted in films with exceptional tensile strength, toughness, UV-shielding properties, and reduced hydrophilicity, making them suitable for industrial packaging and UV shielding [61].

4. Economic implications

The economic viability of producing bioplastics from waste hinges on several factors, including production costs, market prices, and the processes' scalability.

Poly(3-hydroxybutyrate-co-3-hydroxyvalerate) (PHBV) production from low-cost substrates like methane and valeric acid derived from organic waste can reduce costs significantly, with the minimum selling price (MSP) dropping to 8.6 €/kg when biomass concentration is optimized [62]. Similarly, polylactic acid (PLA) produced from corn stover, an agricultural byproduct, shows cost competitiveness with corn grain-based PLA, although higher fixed costs remain a challenge. Upscaling and consumer awareness about environmental benefits are crucial for further cost reductions [63].

For polyethylene furanoate (PEF) and polytrimethylene terephthalate (PTT), derived from PET upcycling, the MSP can be reduced to 1.61 USD/kg with increased recycling rates, demonstrating the economic benefits of a circular plastics economy [64]. Cyanobacterial production of poly(3-hydroxybutyrate) (PHB) using atmospheric CO2 as a carbon source shows potential for cost reduction, but current MSPs are still high at USD 7704 per ton, even with optimized scenarios [65]. The recycling of PET/OPP laminated waste film, achieving a cost–benefit ratio (BCR) of 1.32 at a 30% recycling rate, indicates economic viability, especially with anticipated increases in government support [66].

The microbial production of polyhydroxyalkanoate (PHA) using engineered *Halomonas cupida* J9 and inexpensive feedstocks like corn straw hydrolysate shows promise, with yields reaching 12.57 g/L PHA in unsterile conditions, highlighting the potential for low-cost production from lignocellulose-rich agricultural waste [8, 67]. Conversely, the pyrolysis of high-density polyethylene (HDPE) waste into liquid fuels is currently not economically viable due to high production costs and operational expenditures.

However, it may be more feasible in developing countries with lower utility and labor costs [68]. Overall, while the production of bioplastics from waste shows significant promise, achieving economic viability requires optimizing production processes, increasing recycling rates, and leveraging government support and consumer awareness to drive market demand.

5. Challenges and opportunities

Scaling the production of bioplastics from lab to industrial scale presents several technical challenges. One significant barrier is the high cost and complexity of pretreatment and catalyst refurbishment in synthesizing poly(ethylene furanoate) (PEF), a next-generation bio-based plastic. Efficient substrate utilization and pathway optimization are crucial for the microbial-based biorefinery processes used to produce PEF monomers, which require advanced bioengineering and bioprocessing strategies to be viable at an industrial scale.

Similarly, producing polyhydroxyalkanoate (PHA) from activated sludge faces challenges in optimizing the enrichment process and feast/famine ratio to enhance productivity and product quality. The selection of appropriate extraction methods and integration of nitrogen removal are also critical to reducing costs and improving scalability [69]. Additionally, the production of bioplastics from organic waste, such as microalgae and agro-industrial waste, is hindered by low production yields and high costs. The lack of facilities and the need for costly feedstocks further complicate the industrial-scale deployment of these bioplastics [70].

Despite their potential, compostable bioplastics face issues with disintegration rates in industrial composting plants, where the composition and heterogeneity of the composting substrate significantly affect degradation rates. This discrepancy between lab-scale and industrial-scale results underscores the need for tailored composting conditions to ensure effective biodegradation [71]. Moreover, the management of compostable bioplastics in waste treatment systems is problematic due to their poor degradability under anaerobic conditions and the technical challenges in separating them from conventional plastics during pretreatment processes. This leads to increased process residues and lower biogas production, highlighting the need for improved waste management strategies and better recognition of compostable products. Addressing these technical barriers is essential for the successful industrial-scale production and commercialization of bioplastics.

Several factors, including environmental benefits, consumer awareness, and economic considerations, influence bioplastics' market dynamics and consumer perception. Bioplastics decomposed by microorganisms into eco-friendly byproducts are gaining traction due to their potential to alleviate the energy crisis and reduce dependence on nonrenewable resources, with an annual growth rate of 10–20% [18].

Consumer acceptance is crucial for the success of bioplastics, as highlighted by studies showing that attitudes and values significantly impact consumer behavior, while socio-demographics have minimal effect [72]. Educating consumers and raising awareness through labels and certifications are recommended strategies to enhance acceptance [72]. In Zimbabwe, self-identity, perceived value, cognitive biases, and self-congruity positively influence consumer acceptance of bioplastics. This suggests that marketing efforts should focus on these aspects to facilitate the transition from a linear to a circular economic system [73].

Additionally, adopting green technologies to reduce production costs can increase the perceived value of bioplastics [73]. Edible food packaging, an innovative application of bioplastics, has shown promising consumer acceptance, with a significant percentage of consumers willing to purchase edible products, indicating a positive attitude towards sustainable alternatives to single-use plastics [73]. Overall, the market dynamics for bioplastics are favorable, with growing consumer acceptance driven by environmental consciousness and effective marketing strategies that highlight the benefits and value of bioplastic products.

6. Policy and regulatory framework

Governmental and international policies support waste-to-wealth initiatives by providing the necessary regulatory framework, economic incentives, and public awareness campaigns to drive sustainable waste management practices. Policies such as extended producer responsibility, environmental taxes, and corporate sustainability practices are crucial for operationalizing the circular economy and transitioning towards sustainable development [74]. Governmental support is also essential in establishing sustainable supply chains incorporating waste recycling, as seen in the structure involving the government, manufacturers, suppliers, waste depots, and recyclers. These highlight supportive policies' significant role in resource conservation and energy storage [75].

Additionally, policy support substantially positively affects waste classification behavior and effectiveness, particularly in rural areas, by enhancing environmental protection perceptions among residents and narrowing regional environmental governance gaps [76]. Universities and other institutions can benefit from governmental policies that address operational challenges in bio-waste management, thereby turning waste into a potential source of institutional income and reducing burdensome operational costs [77].

Furthermore, international policies that promote recycling, reuse, and energy recovery are essential in an integrated approach towards waste management, ensuring that even lower-income countries can benefit from advanced waste management technologies and avoid environmental disasters [74]. Collectively, these policies facilitate the transformation of waste into valuable resources and contribute to broader environmental and economic sustainability goals.

7. Prospects and integration into the circular economy

Bioplastics have significant potential to integrate into and enhance existing circular economy models by addressing both environmental and economic challenges associated with conventional plastics. Bioplastics, which are bio-based and/or biodegradable, can reduce greenhouse gas emissions and mitigate plastic waste issues markedly when removal is difficult [78].

The use of sustainable materials such as polyhydroxyalkanoates (PHAs) and polylactic acid (PLA) derived from algae and other bio-feedstocks offers an environmentally friendly alternative to petroleum-based plastics, which is crucial given the escalating environmental concerns [79]. However, integrating bioplastics into the circular economy has challenges, including high production costs and the need for effective downstream recycling routes, such as mechanical, chemical, or biological recycling [78, 80].

The application of advanced technologies like the Internet of Things (IoT) and machine learning can optimize bioplastic production processes, enhancing efficiency and output and thus supporting a more sustainable production model [79]. Additionally, using terahertz (THz) systems for nondestructive testing and quality control can improve bioplastics' economic productivity and environmental impact by ensuring high-quality production and reducing waste [81].

To overcome barriers such as insufficient resources and resistance to change, a multifaceted approach involving increased financial and technical support, enhanced stakeholder engagement, and promoting sustainable consumption practices is essential [79]. Moreover, developing a strategic and comprehensive bioplastic production and integration plan, including policy support and incentives, is crucial for scaling up and achieving cost competitiveness [80]. By addressing these challenges and leveraging advanced technologies, bioplastics can significantly contribute to a sustainable circular economy, reducing reliance on petrochemical feedstocks and minimizing environmental degradation [78–81].

8. Conclusion

The chapter explores the development and use of bioplastics, which are derived from renewable resources and naturally decompose with the help of microorganisms, breaking down into carbon dioxide, water, organic substances, and inorganic compounds. These innovative materials represent a sustainable alternative to traditional plastics, which have dominated since the mid-nineteenth century when semisynthetic plastics like Parkesine and celluloid were first synthesized. The widespread adoption of synthetic polymers was driven by their desirable properties, such as high molecular weight, lightweight, and durability.

However, despite these benefits, synthetic plastics have led to significant environmental challenges due to their low degradation rates, contributing to pollution in oceans and other ecosystems. This growing concern has spurred interest in bioplastics, categorized into two main classes: agro-polymers, including starch, proteins, and chitin, and biopolyesters, such as polyhydroxyalkanoates and polylactic acid.

A notable breakthrough in the field has been the development of bioplastics incorporating lignin-Cu nanoparticles (LNPCu) into poly(vinyl alcohol) (PVA). This innovation mimics the granular nanostructure in mussel byssus cuticles, resulting in films with enhanced mechanical properties and UV-shielding capabilities. Initially, bioplastic production focused on polylactic acid (PLA) but has since expanded to include other materials like polysaccharides, polyhydroxyalkanoates (PHAs), and polyhydroxy butyrates (PHBs) obtained from bacterial processes.

The potential of using agro-industrial waste, such as crop residues and byproducts from sugarcane processing, to produce bioplastics is also highlighted. This approach not only helps manage waste but also promotes sustainability. Agricultural wastes like corn stalks, wheat straw, and sugarcane bagasse, rich in lignocellulosic polymers, are particularly suitable for conversion into bioplastics through microbial processes.

However, the economic viability of bioplastics is influenced by factors such as production costs, market prices, and scalability. Efforts to reduce costs include using low-cost substrates and optimizing biomass concentration. As demand increases and oil prices rise, the commercial manufacturing capacity of bioplastics is expected to grow significantly, with Europe likely to hold a substantial share of the market.

Finally, consumer acceptance plays a crucial role in the success of bioplastics, with studies indicating that attitudes and values significantly impact behavior, while sociodemographic factors have minimal effect. Governmental and international policies are also vital, providing regulatory frameworks, economic incentives, and public awareness campaigns to support waste-to-wealth initiatives and drive sustainable waste management practices.

Acknowledgements

The authors are grateful for the support from CONAHYT, México.

EFH and RAGS authors thank the support for COFAA-IPN and EDI-IPN.

Conflict of interest

The authors declare no conflict of interest.

DOI: http://dx.doi.org/10.5772/intechopen.1006837

Author details

Elizabeta Hernández Domínguez[1], María de la Luz Sánchez Mundo[2], Rosalía América González Soto[3] and Emmanuel Flores Huicochea[3*]

1 Tecnológico Nacional de México, Instituto Tecnológico Superior de Acayucan, Carretera Costera del Golfo km, Acayucan, Veracruz, Mexico

2 Tecnológico Nacional de México, Instituto Tecnológico Superior de las Choapas, Las Choapas, Veracruz, Mexico

3 Instituto Politécnico Nacional, Centro de Desarrollo de Productos Bióticos, Yautepec, Mexico

*Address all correspondence to: efloresh@ipn.mx

References

[1] Rasmussen SC. From Parkesine to celluloid: The birth of organic plastics. Angewandte Chemie – International Edition. 2021;**60**:8012-8016. DOI: 10.1002/anie.202015095

[2] Jansen JA. Plastics – It's all about molecular structure. Plastics Engineering. 2016;**72**:44-49. DOI: 10.1002/j.1941-9635.2016.tb01587.x

[3] Brinson HF, Brinson LC. Characteristics, applications and properties of polymers. In: Polymer Engineering Science and Viscoelasticity. US: Springer; 2015. pp. 57-100. DOI: 10.1007/978-1-4899-7485-3_3

[4] Zhang Y, Yu X, Cheng Z. Research on the application of synthetic polymer materials in contemporary public art. Polymers. 2022;**14**:1208. DOI: 10.3390/polym14061208

[5] Ibrahim ID, Jamiru T, Sadiku ER, Hamam A, Kupolati WK. Applications of polymers in the biomedical field. Current Trends in Biomedical Engineering & Biosciences. 2017;**4**:102-104. DOI: 10.19080/ctbeb.2017.04.555650

[6] Pan Y, Farmahini-Farahani M, O'Hearn P, et al. An overview of bio-based polymers for packaging materials. Journal of Bioresources and Bioproducts. 1996;**1**(3):106-113. DOI: 10.21967/jbb.v1i3.49

[7] Caruso MM, Davis DA, Shen Q, et al. Mechanically-induced chemical changes in polymeric materials. Chemical Reviews. 2009;**109**:5755-5798. DOI: 10.1021/cr9001353

[8] Tsuge Y, Kawaguchi H, Sasaki K, et al. Engineering cell factories for producing building block chemicals for bio-polymer synthesis. Microbial Cell Factories. 2016;**15**:19. DOI: 10.1186/s12934-016-0411-0

[9] Lebreton LCM, van der Zwet J, Damsteeg J-W, et al. River plastic emissions to the world's oceans. Nature Communications. 2017;**8**:15611

[10] Geyer R, Jambeck J, Law KL. Production, use, and fate of all plastics ever made. Science Advances. 2017;**3**:3

[11] Gahleitner MA. A second life for polymers... Needs more research. Express Polymer Letters. 2016;**10**:187

[12] Worm B, Lotze H, Jubinville I, et al. Plastic as a persistent marine pollutant. Annual Review of Environment and Resources. 2017;**42**:1-26

[13] Plastics Europe. Plastics – The Fast Facts 2023. 2023. Available from: https://plasticseurope.org/wp-content/uploads/2023/03/Plastics-the-Facts-2023.pdf

[14] Jâms IB, Windsor FM, Poudevigne-Durance T, et al. Estimating the size distribution of plastics ingested by animals. Nature Communications. 2020;**11**:1594

[15] Hussain A, Lin C, Nguyen MK. Biodegradation of different types of bioplastics through composting – A recent trend in green recycling. Cataysts. 2023;**13**:294

[16] BioCycle. Composting Process Conditions And Bioplastic Degradation Rates. 2023. Available from: https://www.biocycle.net/composting-process-conditions-and-bioplastic-degradation-rates

[17] Ostle C, Thompson RC, Broughton D, et al. The rise in ocean plastics evidenced from a 60-year time series. Nature Communications. 2019;**10**:1622

[18] Nisar B, Pahalvi HN, Gulzar A, et al. Chapter 25 – Bioplastics: Solution to a green environment and sustainability. In: Srivastav AL, Grewal AS, Markandeya, et al., editors. Advances in Pollution Research. Cambridge, MA, United States: Elsevier; 2024. pp. 261-269

[19] Jabeen M, Tarıq K, Hussain SU. Bioplastic an alternative to plastic in modern world: A systemized review. Environmental Research and Technology. 2024. Available from: https://api.semanticscholar.org/CorpusID:269844945

[20] Ali S, Isha I, Chang YC. Ecotoxicological impact of bioplastics biodegradation: A comprehensive review. PRO. 2023;**11**:3345-3470. DOI: 10.20944/preprints202311.1858.v1

[21] BioPoly Lab. A Short History of Bioplastics. 2020. Available from: https://biopolylab.com/2020/06/a-short-history-of-bioplastics

[22] Barret A. History of Bioplastics. Bioplastics News; 2018. Available from: https://bioplasticsnews.com/2018/07/05/history-of-bioplastics/

[23] Narancic T, Cerrone F, Beagan N, O'Connor KE. Recent advances in bioplastics: Application and biodegradation. Polymers. 2020;**12**:920. DOI: 10.3390/polym12040920

[24] Chauhan K, Kaur R, Chauhan I. Sustainable bioplastic: A comprehensive review on sources, methods, advantages, and applications of bioplastics. Polymer-Plastics Technology and Materials. 2024;**63**:913-938

[25] Naik B, Kumar V, Rizwanuddin S, et al. Agro-industrial waste: A cost-effective and eco-friendly substrate to produce amylase. Food Production, Processing and Nutrition. 2023;**5**:1-12

[26] Anal A, Panesar P, Kaur R. Agro-industrial waste as wealth. In: Valorization of Agro-Industrial Byproucts. 2022. pp. 1-10

[27] Bhattacharya I, Sahiba M, Suhag M. Sustainable agricultural waste management practices and approaches: An overview. Futuristic Trends in Agriculture Engineering and Food. 2024;**3**:156-166

[28] Alamsyah A. Agro-industrial waste enzymes: Perspectives in circular economy. Current Opinion in Green and Sustainable Chemistry. 2022;**34**:100585

[29] Dwivedi S, Tanveer A, Yadav S, et al. Agro-wastes for cost effective production of industrially important microbial enzymes. In: Microbial Biotechnology: Role in Ecological Sustainability and Research. NJ, USA: Wiley; 2022. pp. 435-460

[30] Jardine PE, Gosling WD, Lomax BH, et al. Chemotaxonomy of domesticated grasses: A pathway to understanding the origins of agriculture. Journal of Micropalaeontology. 2019;**38**:83-95

[31] FAOSTAT. Crops and Livestock Products. 2024. Available from: https://www.fao.org/faostat/

[32] Suryaningrum LH, Samsudin R. Improvement quality of sugar cane bagasse as fish feed ingredient. IOP Conference Series: Earth and Environmental Science. 2021;**679**:12003

[33] Antonio Bizzo W, Lenço PC, Carvalho DJ, et al. The generation of residual biomass during the production

of bio-ethanol from sugarcane, its characterization and its use in energy production. Renewable and Sustainable Energy Reviews. 2014;**29**:589-603

[34] Nath A, Das K, Dhal GC. Global status of agricultural waste-based industries, challenges, and future prospects. In: Neelancherry R, Gao B, Wisniewski A Jr, editors. Agricultural Waste to Value-Added Products. Singapore: Springer; 2023. DOI: 10.1007/978-981-99-4472-9_2

[35] Raungrusmee S. Carbohydrate-Based Agro-Industrial Waste;2022. pp. 185-200

[36] Siva, Janika A, Afrrin M, Rahul R, et al. Valorization of sugarcane bagasse in food packaging. In: Kajla SPBP, editor. Agro-Wastes for Packaging Applications. Boca Ratón, Florida, USA: CRC; 2024. pp. 169-196

[37] Hossam Y, Fahim I. Towards a circular economy: Fabrication and characterization of biodegradable plates from sugarcane waste. Frontiers in Sustainable Food Systems. 2023;**7**:1-13. DOI: 10.3389/fsufs.2023.1220324

[38] Li Y, Chai J, Wang R, et al. Utilization of sugarcane bagasse ash (SCBA) in construction technology: A state-of-the-art review. Journal of Building Engineering. 2022;**56**:104774

[39] Martillo Aseffe JA, Lesme Jaén R, Oliva Ruíz LO. Estimación del potencial energético de la tusa en la provincia de los ríos y guayas, Ecuador. Centro Azúcar. 2020;**47**:11-21

[40] Gupte AP, Basaglia M, Casella S, et al. Rice waste streams as a promising source of biofuels: Feedstocks, biotechnologies and future perspectives. Renewable and Sustainable Energy Reviews. 2022;**167**:112673

[41] Urrea-Ceferino G, Mojica M. Crop waste management proposal in rice systems at the department of Cordoba, Colombia. Economia Agro-Alimentare. 2023;**25**:167-189

[42] Chaturvedi S, Verma A, Sethi SK, et al. Chapter 15 – Stalk fibers (rice, wheat, barley, etc.) composites and applications. In: Mavinkere Rangappa S, Parameswaranpillai J, Siengchin S, et al., editors. The Textile Institute Book Series. Cambridge, MA, United States: Woodhead Publishing; 2022. pp. 347-362

[43] Hoskinson R, Thompson D, Foust T, et al. Distributed Fungal Harvest of Higher Value Wheat Straw Components. 2002. DOI: 10.13031/2013.9714

[44] Triveni D, Uma Jyothi K, Dorajee Rao AVD, et al. Correlation and path analysis for yield and yield contributing traits in bitter gourd (Momordica charantia L.). Vegetos. 2021;**34**:944-950

[45] Warner RE, Havera SP, David LM. Effects of autumn tillage systems on corn and soybean harvest residues in Illinois. Journal of Wildlife Management. 1985;**49**:185-190

[46] Dwibedi S, Mohanty MK, Pandey V, et al. Sustainable Biowaste Management in Cereal Systems: A Review2021. DOI: 10.5772/intechopen.97308

[47] Kamm B, Kamm M. Biorefineries—Multi product processes. Advances in Biochemical Engineering/Biotechnology. 2007;**105**:175-204

[48] Pitt J, Heinemeyer C. Environment, Ethics and Cultures. Introducing ideas of a circular economy BT – Environment, ethics and cultures: Design and technology Education's contribution to sustainable global futures. In: Stables K, Keirl S, editors. Rotterdam: SensePublishers. 2015. pp. 245-260

[49] Murray A, Skene K, Haynes K. The circular economy: An interdisciplinary exploration of the concept and application in a global context. Journal of Business Ethics. 2017;**140**:369-380

[50] Andriani Y, Rahma C, Pratama R. Exploring the transformation of food waste into bioplastic materials: A review. Asian Journal of Biotechnology and Bioresource Technology. 2024;**10**:1-11. DOI: 10.9734/ajb2t/2024/v10i1193

[51] Biswas A, Pal S, Mitra RD. A review on- preparation of bio plastics from Agri-waste. The Bioscan. 2024;**19**:9-12

[52] Narancic T, Cerrone F, Beagan N, O'Connor KE. Recent advances in bioplastics: Application and biodegradation. Polymers. 2023;**12**(920):167243-167258. DOI: 10.3390/polym12040920

[53] Francis D, Parayil DJ. Microbial Technology in Bioplastic Production and Engineering. In: Handbook of Bioplastics and Biocomposites Engineering Applications. Beverly, MA, USA: John Wiley & Sons, Inc.; 2022. pp. 121-148

[54] Shynzhyrbai K, Azat S, Mataev M, et al. Development of technology for obtaining a biodegradable polymer. Vestn NÂC RK. 2023;**3**:72-80. DOI: 10.52676/1729-7885-2023-3-72-80

[55] Omar MN, Minggu MM, Muhammad NAN, et al. Towards consolidated bioprocessing of biomass and plastic substrates for semisynthetic production of bio-poly(ethylene furanoate) (PEF) polymer using omics-guided construction of artificial microbial consortia. Enzyme and Microbial Technology. 2024;**904**:167243-167258. DOI: 10.1016/j.enzmictec.2024.110429

[56] Al-Khairy D, Fu W, Alzahmi A, et al. Closing the gap between bio-based and petroleum-based plastic through bioengineering. Microorganisms. 2022;**10**:2320

[57] Shahid MA, Akter S, Ferdous J, et al. Cellulose and starch-based bioplastics: A review of advances and challenges for sustainability. Polymer-Plastics Technology and Materials. 2024;**10**:2330-2344. DOI: 10.1080/25740881.2024.2329980

[58] Hossain MT, Shahid MA, Akter S, Ferdous J, Afroz K, Refat KRI, et al. Cellulose and starch-based bioplastics: A review of advances and challenges for sustainability. Polymer-Plastics Technology and Materials. 2024;**63**(10):1329-1349. DOI: 10.1080/25740881.2024.2329980

[59] Acosta DJ, Alper HS. Advances in enzymatic and organismal technologies for the recycling and upcycling of petroleum-derived plastic waste. Current Opinion in Biotechnology. 2023;**84**:103021

[60] Rahmati F, Sethi D, Shu W, et al. Advances in microbial exoenzymes bioengineering for improvement of bioplastics degradation. Chemosphere. 2024;**355**:141749-141767

[61] Rahmati F, Sethi D, Shu W, et al. Advances in microbial exoenzymes bioengineering for improvement of bioplastics degradation. Chemosphere. 2024;**355**:141749-141767

[62] Amabile C, Abate T, Muñoz R, et al. Techno-economic assessment of biopolymer production from methane and volatile fatty acids: Effect of the reactor size and biomass concentration on the poly(3-hydroxybutyrate-co-3-hydroxyvalerate) selling price. Science of the Total Environment. 2024;**929**:172599

[63] Wellenreuther C, Wolf A, Zander N. Cost competitiveness of sustainable

bioplastic feedstocks – A Monte Carlo analysis for polylactic acid. Cleaner Engineering and Technology. 2022;**6**:100411

[64] Roux M, Varrone C. Assessing the economic viability of the plastic biorefinery concept and its contribution to a more circular plastic sector. Polymers (Basel). 2021;**13**:3883

[65] Price S, Kuzhiumparambil U, Pernice M, et al. Techno-economic analysis of cyanobacterial PHB bioplastic production. Journal of Environmental Chemical Engineering. 2022;**10**:107502

[66] Park MS, Kim DY, Yang SJ, et al. An economic analysis study of recycling PET·OPP laminated film waste generated during DECO film manufacturing. Resource Recycling. 2023;**32**:57-67

[67] Wang S, Liu Y, Guo H, et al. Establishment of low-cost production platforms of polyhydroxyalkanoate bioplastics from Halomonas cupida J9. Biotechnology and Bioengineering. 2024;**121**:2106-2120

[68] Zein SH, Grogan CT, Yansaneh OY, et al. Pyrolysis of high-density polyethylene waste plastic to liquid fuels – Modelling and economic analysis. PRO. 2022;**10**:1503

[69] Zhang Z, Wang Y, Wang X, et al. Towards scaling-up implementation of polyhydroxyalkanoate (PHA) production from activated sludge: Progress and challenges. Journal of Cleaner Production. 2024;**447**:141542

[70] Ali Z, Abdullah M, Yasin MT, Amanat K, Ahmad K, Ahmed I. Organic waste-to-bioplastics: Conversion with eco-friendly technologies and approaches for sustainable environment. Environmental Research. 2024;**244**:117949-117958. DOI: 10.1016/j.envres.2023. [Epub 2023 Dec 17]

[71] Ali Z, Abdullah M, Yasin MT, Amanat K, Ahmad K, Ahmed I, et al. Organic waste-to-bioplastics: Conversion with eco-friendly technologies and approaches for sustainable environment. Environmental Research. 2024;**244**:117949-117958. DOI: 10.1016/j.envres.2023.117949. Epub 2023 Dec 17

[72] Chong ZK, Hofmann A, Haye M, Wilson S, Sohoo I, Alassali A, et al. Lab-scale and full-scale industrial composting of biodegradable plastic blends for packaging. Open Research Europe. 2024;**2**:101-134. DOI: 10.12688/openreseurope.14893.3

[73] Emmanuel M, Joe M. Factors that influence the successful marketing of bioplastic products in Zimbabwe: Towards a circular economy by 2030. International Journal of Innovative Research and Scientific Studies. 2023;**6**:389-398

[74] Chenavaz RY, Dimitrov S. From waste to wealth: Policies to promote the circular economy. Journal of Cleaner Production. 2024;**443**:141086

[75] Jafari H, Safarzadeh S. Effects of governmental supportive policies on waste management for two substitutable products made of virgin and waste materials: A game-theoretic approach. Waste Management & Research: The Journal for a Sustainable Circular Economy. 2024;**443**:141086-144096. DOI: 10.1177/0734242X241231399

[76] Huang Y, Zhong Z. How does policy support affect the behavior and effectiveness of domestic waste classification? The mediating role of environmental protection perception. International Journal of Environmental

Research and Public Health. 2023;**20**. DOI: 10.3390/ijerph20032427

[77] Kamsano NS, Sohaili J. Waste-to-wealth: Bio-recycling Centre as living laboratory element to create integrated bio-waste Management for Institution. International Journal of Waste Resources. 2016;**6**:2427-2442. DOI: 10.4172/2252-5211.1000217

[78] Serrano-Aguirre L, Prieto MA. Can bioplastics always offer a truly sustainable alternative to fossil-based plastics? Microbial Biotechnology. 2024;**17**:1000217-1000220. DOI: 10.1111/1751-7915.14458

[79] Bin Abu Sofian ADA, Lim HR, Manickam S, et al. Towards a sustainable circular economy: Algae-based bioplastics and the role of internet-of-things and machine learning. ChemBioEng Reviews. 2024;**11**:39-59

[80] Shah KU, Gangadeen I. Integrating bioplastics into the US plastics supply chain: Towards a policy research agenda for the bioplastic transition. Frontiers in Environmental Science. 2023;**11**:39-59. DOI: 10.3389/fenvs.2023.1245846

[81] Abina A, Korošec T, Puc U, et al. Review of bioplastics characterisation by terahertz techniques in the view of ensuring a circular economy. Photonics. 2023;**10**:883

Chapter 3

Bioplastic in Water Streams Challenges

Mohamed Said Mahmoud

Abstract

The idea of bioplastics might sound very commendable, being a replacement for traditional plastics. However, it is actually a rather intricate problem when it comes to their impact on water bodies. While biodegradability is considered a key asset, high temperatures, controlled humidity, and other optimum conditions required for decomposition are seldom achieved in marine and freshwater ecosystems. Hence, most bioplastics, especially PLA, are conveyed by water streams, contributing to microplastic pollution, disrupting the processes of wastewater treatments, and causing direct harm to the aquatic fauna due to entanglement and ingestion. This chapter elaborates on these different challenges underlined by the gap between perceived and actual degradability in water bodies. On the other hand, it is also pointing toward some promising solutions, such as developing marine-degradable bioplastics like PHA, new technologies in wastewater treatment, and policy and regulation playing an important role in fostering responsible production and consumption with disposal. Above all, finding full value in bioplastics while limiting unintended impacts on aquatic ecosystems demands a series of measures involving scientific innovation, policy reform, and greater public awareness.

Keywords: bioplastics, biodegradability, aquatic environments, microplastics, marine pollution, sustainable development, circular bioeconomy

1. Introduction

Bioplastics offer hope as a replacement for ordinary plastics, though the complexity of their impact on water bodies is found. Despite its claimed advantage of biodegradability, the ocean and river environments used for disposal do not tend to have the high/debris processing temperatures and regulated moisture levels required for decomposition. As a consequence, many of the bioplastics, but especially PLA can be found in water streams and add to microplastic pollution, disturb wastewater treatment processes, or cause direct harm to aquatic fauna by entwining or ingestion. An illustration of this type of problem is discussed in this chapter: the difference between perceived and actual biodegradability in aquatic media. It also investigates potential solutions such as the advancement of marine-degradable bioplastics, including polyhydroxyalkanoates (PHA), progress in wastewater treatment technologies, and the importance of policy and regulation as well to build toward a circular production-consumption-disposal model. In the end, achieving complete.

2. Bioplastics in water: A summary of challenges and solutions

Bioplastics are produced from renewable feedstocks like corn starch and are often promoted as a realistic substitute for conventional oil-based plastics. Of the traits that have captured consumer and media fancy, their biodegradability-natural breaking down-is one. However, as this chapter illustrates, the biodegradability of many bioplastics, and in particular PLA, is actually quite conditional. They require very specific environments, like industrial composting facilities with high temperatures and controlled humidity levels, in order to decompose efficiently [1]. All these factors are absent in the case of aquatic water bodies like oceans, rivers, and lakes; hence, the bioplastic debris persists and gets as cumbersome as traditional plastics. This persistence has several significant implications:

- Microplastic pollution: Bioplastics decompose into microplastics-small particles that are consumed by aquatic organisms. The ingestion can result in physical harm, like blockages in the digestive system, or result in the bioaccumulation of toxins absorbed by the microplastics and pose a threat to the whole food web, including humans consuming seafood.

- Wastewater treatment challenges: Bioplastics of variable rates of biodegradability can disrupt the wastewater treatment process. Some degrade too quickly, overloading the biogas capture mechanisms, while others resist decomposition and inhibit sludge treatment, with the possible release of microplastics into the treated water released into the environment.

- Direct harm to marine life: Direct harm to marine life: The larger size bioplastic debris, which is similar in size to conventional plastics, can entangle marine animals, and ingestion causes blockages in the digestive system, possibly affecting the aquatic ecosystem.

Despite these realities, the chapter also provides a potential source of hope based on continuing research and development:

- New bioplastics for the marine environment: Bioplastics, like PHA (polyhydroxyalkanoates), are under development specifically to biodegrade under marine conditions.

- Advanced Waste Treatment Technologies: Anaerobic digestion and filtration systems are good for improving waste treatment at WTPs with regard to bioplastic wastes, allowing the avoidance of microplastic discharge altogether.

- Policy and Regulation: Harmonized labeling with regards to biodegradability, extended producers' responsibility, and international coordination give a framework within which bioplastics are manufactured responsibly and used, with proper disposal.

Conclusively, the chapter shows that scientific advances, policy change, and an increase in public awareness will be needed to tap the full potential of bioplastics as a viable alternative, with minimized negative impacts on aquatic environments.

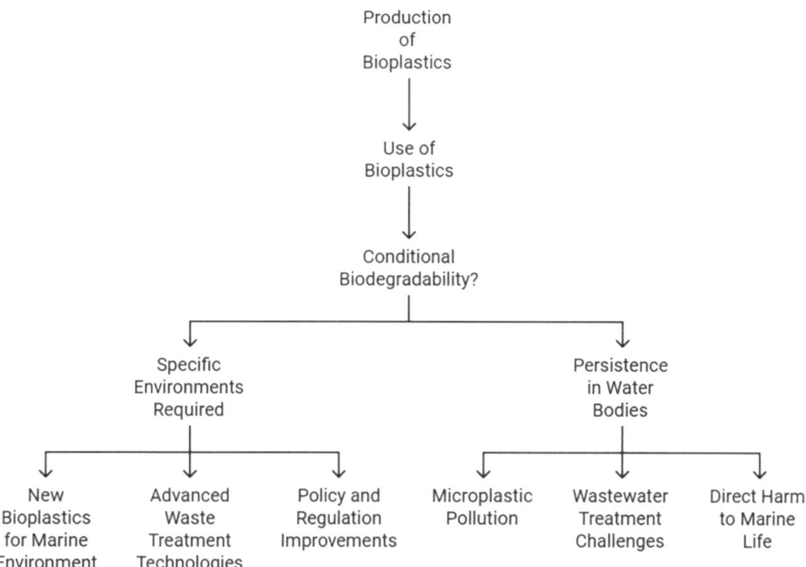

Figure 1.
Bioplastics in water: A summary of challenges and solutions.

The general trend is one of guarded optimism, noting that though there are challenges, indeed solutions lie within reach through collaboration of effort and further innovation (**Figure 1**).

3. Unraveling the bioplastics dilemma at aquatic bodies

Plastic pollution has been an old refrain in the twenty-first century, from images of swirling plastic gyres around ocean currents to pictures of marine animals tangled in cast-off fishing nets the cry has gone out around the world for sustainable alternatives. Bioplastics entered the fray, literally hewn from renewable resources such as corn starch or sugarcane, and promised a solution to the otherwise intractable problem of plastic waste. But the story of bioplastics is anything but a straightforward success [2, 3]. While being lauded for their biodegradability and much-reduced reliance on fossil fuels, it is with respect to aquatic environments that the impact of such bioplastics has generated perhaps one of the most complex challenges often overlooked. This chapter navigates the stormy waters of bioplastics in water streams, highlighting diverse problems that they create and laying out the right course for a more sustainable future. The Paradox of Biodegradability: Myth vs. Reality.

The main attractiveness of bioplastics is the badly proclaimed biodegradability. The very denomination invokes the idea of materials returning to nature so softly that no signs of them will remain. Reality is different, as the degradation of bioplastics is nuanced and highly dependent on environmental conditions. Though indeed some bioplastics can degrade fast in very peculiar conditions in industrial composting facilities, where temperature and humidity are kept under control, Innocenti et al. [4] their destiny in aquatic ecosystems follows a different path.

The currently used bioplastics, especially PLA - polylactic acid, are supposed to be degraded under very controlled conditions, including high temperatures of 50–70°C

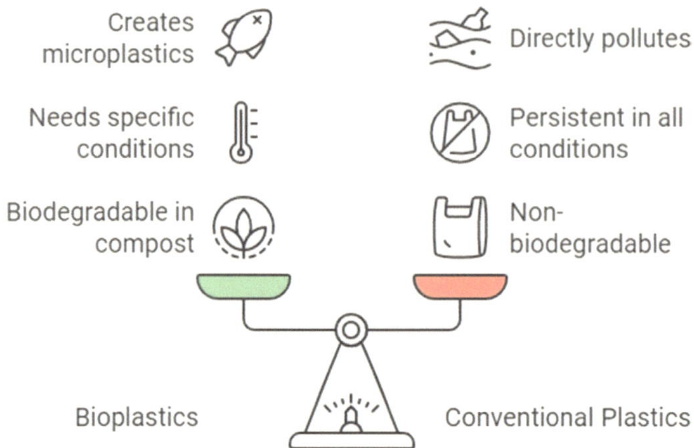

Figure 2.
Evaluating the environmental impact of bioplastics vs. conventional plastics.

and constant moisture. These conditions are hardly ever reached in normal aquatic environments; therefore, the temperatures remain rather low, as does the availability of oxygen. Consequently, bioplastics can remain longer in water streams than expected, thereby acquiring a conventional plastic-like longevity. This persistence undermines the core environmental benefit of bioplastics and perpetuates the growing problem of microplastic pollution.

Besides that, the term "biodegradable" is so ill-defined, and, in many cases, it has been misused. Vagueness in its definition creates confusion among consumers and gives way to many problems in handling and managing waste. A product labeled as "biodegradable" might easily degrade in a compost bin but remain in the ocean for decades. This more or less points to the critical need for standardized labeling and certification schemes, which ensure that the material is biodegradable under various environmental conditions. While these two, among several others like the European "OK compost" and the American "BPI Certified Compostable", do bring important benchmarks for industrial composting, they tell very little about their fates in either marine or freshwater systems (**Figure 2**) [3, 5].

4. The microplastic menace: An invisible enemy

As bioplastics stay in the water, they begin to fragment into microplastic pollution, which is omnipresent. Microplastics are small particles of plastic less than 5 mm in size, and they are becoming fast omnipresent in our oceans, rivers, and lakes. Very harmful to aquatic life, some microplastics emanate from the breakdown of larger pieces of plastic, though others, such as the microbeads found in cosmetics, are manufactured at that size purposefully. Bioplastics, although deriving from renewable resources, can also contribute to microplastics and, hence, become part of the already existing load of aquatic ecosystems [4, 6].

The impacts of microplastics on marine organisms are multifaceted and still a subject of active research. Aquatic organisms usually ingest microplastic particles either through the mistaken identification of these particles as food items or *via*

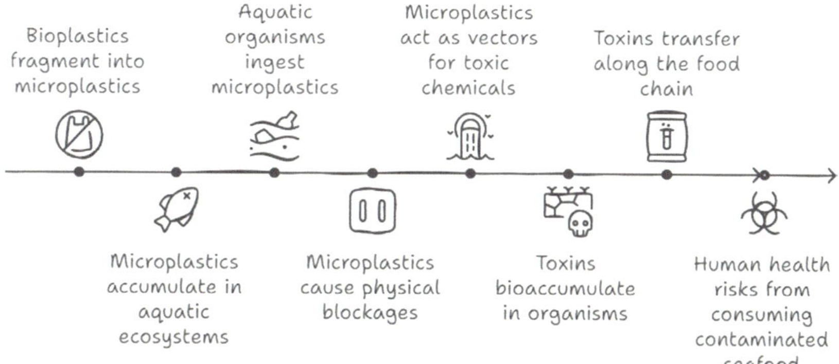

Figure 3.
Bioplastics to microplastics: A hidden pathway to human health risks.

passive ingested food during filter-feeding processes. Particles may be retained within the digestive system, causing physical blockages that could lead to malnutrition or starvation. Moreover, microplastics can act as vectors for toxic chemicals: they absorb from surrounding water POPs and transfer them into the organisms that consume the plastics. This leads to bioaccumulation of toxins which may be transferred along the food chain and finally affect human health upon consumption of seafood (**Figure 3**).

5. Wastewater woes: Treatment plant challenges

All too often, bioplastics make their way through our waste streams to wastewater treatment plants. These plants are critical to removing the pollutants from wastewater before that water is released into the environment. However, bioplastics in the wastewater create new challenges for the operators of those treatment plants.

However, the different biodegradability rates of different types of bioplastics interfere with set treatment processes regarding the efficiency of sludge digestion. Accordingly, Lambert and Wagner [6] explained that some plastics degrade too rapidly and produce so much biogas that exceeds a plant's treating capacity to capture and utilize it, while others that are resistant to degradation may hence start to build up in the sludge and interfere with dewatering. Another major cause for concern is the breaking of bioplastics into microplastics during their wastewater treatment. The conventional treatment methods are largely incapable of capturing microplastics, which are released in the effluents and find their way into environmental bodies (**Figures 4** and 5) [7].

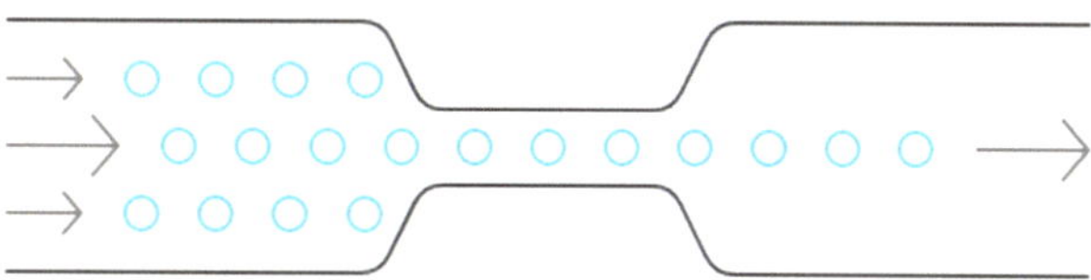

Figure 4.
Disrupts treatment efficiency, leading to environmental contamination risks.

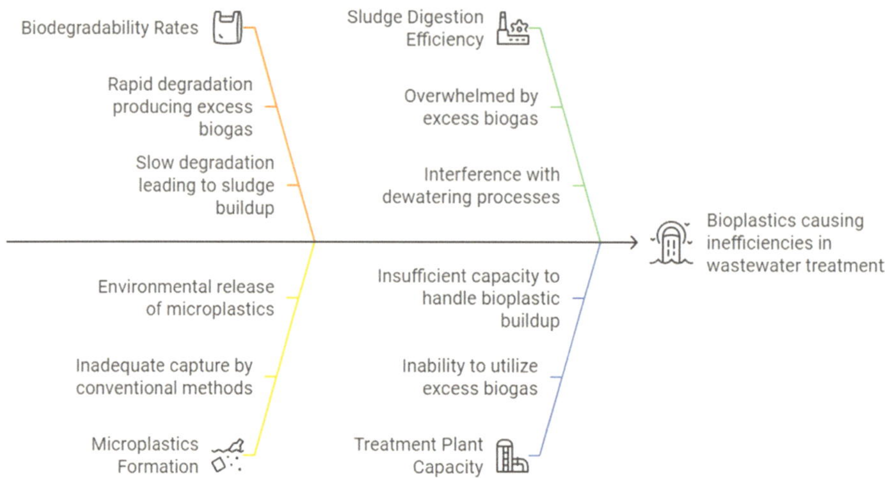

Figure 5.
Challenges of bioplastics in wastewater treatment.

6. The direct impacts on aquatic life include entanglement and ingestion

Besides this insidious threat of microplastics, larger bioplastic debris may cause direct physical damage to aquatic animals through entanglement and ingestion. Although the most frequent attention given is to the entanglement of marine organisms in conventional plastic waste fishing nets and plastic bags, bioplastics also present an entanglement hazard, especially for large animals like sea turtles and marine mammals.

The larger pieces of bioplastic debris can also be ingested by marine animals to their detriment. Similar to conventional plastics, ingested bioplastics block the

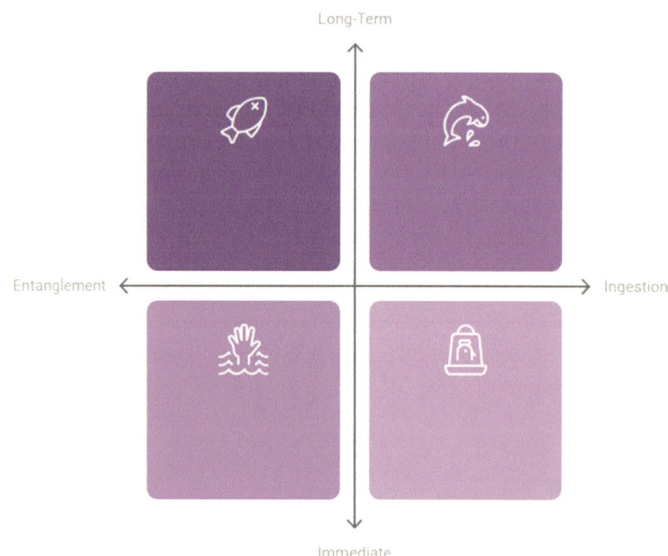

Figure 6.
Impact of bioplastic debris on marine animals.

digestive tract, which can reduce feeding, lead to malnutrition, and potentially cause death. While some bioplastics will eventually break down in the digestive system, this process may be very slow and may still harm the animal during the time period of ingestion (**Figure 6**) [8].

7. Finding our way to solutions: A road to sustainability

However, the story of bioplastics entering water streams is rather one of concern for the environment. It is also a story of steady innovation and adaptation under the urgency of finding sustainable solutions. Researchers are working on new types of bioplastic material, especially those that are biodegradable in a marine environment, PHA, polyhydroxyalkanoates extracted from micro-organisms - Sudesh et al. [9]. Such materials have the potential for complete degradation in marine environments, without producing harmful microplastics or releasing toxic substances.

New anaerobic digestion technologies are being developed that, in the future, have the potential to more effectively treat bioplastic wastes in wastewater treatment facilities [8]. These technologies will make the process of sludge digestion much more effective and increase the produced energy with captured biogas while at the same time minimizing microplastic releases. Innovative filtration systems and other advanced oxidative processes are currently under development for effectively removing microplastics from the effluent of wastewater treatment plants and thus further diminishing their presence in aquatic ecosystems.

But apart from purely technological solutions, much of the key to solving the problem of bioplastics also lies with policy and regulatory frameworks. More transparent labeling and certification schemes would enable consumers to make more informed choices and responsibly dispose of the product. EPR programs may stimulate manufacturers into thinking about and investing in more environmentally friendly bioplastics and methods of waste management at the end-of-life stage. In addition, international collaboration and agreements are necessary with regard to tackling the global spread of plastic pollution and guaranteeing harmonized standards concerning the biodegradability of bioplastics [10].

7.1 Charting a sustainable course: The collaborative approach

The future use of bioplastics in a sustainable society will depend on our ability to address the challenges that they will pose to aquatic environments. This requires a multifaceted approach: integrating technological innovation with policy reform, and increasing public awareness. It involves continuous ecotoxicological research on bioplastics, their after-treatment technologies, and the development of new materials. Such efforts further call for increased transparency and standardization in terms of labeling and certification to enable informed consumer choices [9].

Ultimately, solving the bioplastic challenge is not left to scientists, engineers, or policymakers alone-it is everyone's job. It ranges from manufacturing and retail to consumers and waste management professionals. Embracing the nuance, pursuing innovation, and encouraging collaboration can help map out a more sustainable course for this revolutionary material type, one that would unlock its benefits while minimizing its unintended impacts on our waterways. It is only at this point that we can try to reverse the tide against plastic pollution and protect the health of the waters on this planet [11].

8. Conclusions

The story of bioplastics in our waters offers a neat study of balancing acts. Unique properties of biodegradability make these compounds challengers to conventional plastics, but the persistence in aquatic systems shows that caution and a multiple-point strategy are necessary. This difference in perceptive and real-world biodegradability, together with the associated risks of microplastic generation, impacts on wastewater treatment activities, and external harm to aquatic life demands a predetermined course of intervention.

Long-term answers rest within ongoing research and development. It is partial to an extent though; the breakthroughs of marine-degradable bioplastics such as PHA are a good start toward determining a truly sustainable solution. Technological progress in wastewater treatment, from anaerobic digestion to innovative filtration, could reduce microplastics being released onto land or in the ocean.

Acknowledgements

The author acknowledges the use of the following AI tools in the preparation of this Chapter:

- **Napkin AI** for the creation and optimization of figures and illustrations.

- **DeepSeek Chat** for language polishing and technical refinement of the chapter.

These tools were used in accordance with the journal's **Authorship Policy**, which permits AI assistance provided its contribution is transparently declared. The author assumes full responsibility for the content and interpretation of the work.

Conflict of interest

The authors declare no conflict of interest.

Author details

Mohamed Said Mahmoud
Sanitary and Environmental Institute (SEI), Housing and Building National Research Center (HBRC), Cairo, Egypt

*Address all correspondence to: mphdmicro2012@yahoo.com

References

[1] ElAmin A, editor. Biodegradable Polymers for Food Packaging: Fundamentals and Applications. Springer Science+Business Media; 2007. DOI: 10.1007/978-1-4020-8796-7

[2] Cole M, Lindeque P, Halsband C, Galloway TS. Microplastics as contaminants in the marine environment: A review. Marine Pollution Bulletin. 2011;**62**(12):2588-2597. DOI: 10.1016/j.marpolbul.2011.09.025

[3] Rosenboom J-G, Langer R, Traverso G. The global evolution of bioplastics: Production, markets, and future trends. Current Opinion in Green and Sustainable Chemistry. 2022;**33**:100583. DOI: 10.1016/j.cogsc.2021.100583

[4] Innocenti FD, Razza F, Fieschi M, Bastioli C. Life cycle management in bioplastics production. In: Proceedings of the Third International Conference on Life Cycle Management. Springer; 2007. pp. 1-6. DOI: 10.1007/978-1-4020-9107-5

[5] Gall SC, Thompson RC. The impact of debris on marine life. Marine Pollution Bulletin. 2015;**92**(1-2):170-179. DOI: 10.1016/j.marpolbul.2014.12.041

[6] Lambert S, Wagner M. Environmental performance of bio-based and biodegradable plastics: The road ahead. Chemical Society Reviews. 2017;**46**(22):6855-6871. DOI: 10.1039/C7CS00149E

[7] Nishida H. Recyclable materials: Circulatory recycling system of polylactic acid (PLA). In: ElAmin A, editor. Biodegradable Polymers for Industrial Applications. Woodhead Publishing/Springer; 2007. pp. 343-365. DOI: 10.1007/978-1-4020-9109-9_14

[8] Patel M, Narayan R. How sustainable are biopolymers and biobased products? The hope, the doubt, and the reality. In: Mohanty, AK, Misra M, Drzal LT, editors. Natural Fibers, Biopolymers, and Biocomposites. CRC Press/Taylor & Francis; 2005. pp. 834-851. DOI: 10.1201/9780203508206

[9] Sudesh K, Abe H, Doi Y. Synthesis, structure and properties of polyhydroxyalkanoates: Biological polyesters. Progress in Polymer Science. 2000;**25**(8):1503-1555. DOI: 10.1016/S0079-6700(00)00035-6

[10] Rochman CM, Hoh E, Hentschel BT, Kaye S. Long-term field measurement of sorption of organic contaminants to five types of plastic pellets: Implications for plastic marine debris. Environmental Science & Technology. 2013;**47**(3):1646-1654. DOI: 10.1021/es303700s

[11] Vink ETH, Rábago KR, Glassner DA, Gruber PR. Application of life cycle assessment to NatureWorks™ polylactide (PLA) production. Polymer Degradation and Stability. 2003;**80**(3):403-419. DOI: 10.1016/S0141-3910(02)00372-5

Chapter 4

Bioplastics from Food by-Products

Sungeun Ahn

Abstract

Bioplastics derived from food by-products are emerging as a sustainable alternative to conventional plastics, offering substantial environmental benefits due to their renewable nature and biodegradability. This review explores recent advancements in the development of bioplastics using diverse food by-products, including fruit peels, vegetable waste, and lignocellulosic biomass. It examines the methodologies, results, and implications of various studies; provides in-depth case studies of successful applications; and analyzes production methods concerning their efficiency, cost, and environmental impact. Additionally, this review addresses current market trends, regulatory challenges, and opportunities, proposing future research directions in this rapidly evolving field. High-quality figures, tables, equations, and models are included to provide a detailed understanding of the lifecycle, environmental impact, and market potential of bioplastics. The findings highlight the crucial role of ongoing innovation, regulatory frameworks, and consumer awareness in promoting the widespread adoption of bioplastics derived from food by-products.

Keywords: bioplastics, food by-products, sustainability, food packaging, biodegradability, renewable resources, circular economy, environmental impact, regulatory frameworks

1. Introduction

The extensive use of conventional plastics, particularly single-use plastics, has led to significant environmental problems worldwide. These plastics, primarily derived from nonrenewable fossil fuels, are characterized by their persistence in the environment. Over 300 million tons of plastic waste are generated annually, with a considerable portion ending up in oceans, landfills, and other natural environments, where they can take hundreds to thousands of years to decompose [1]. This accumulation has resulted in severe environmental challenges, such as marine pollution, the formation of microplastics, and the release of harmful chemicals, all of which adversely impact ecosystems and human health [2, 3].

To mitigate these issues, there is an increasing demand for sustainable alternatives that can reduce the reliance on conventional plastics [4]. Bioplastics derived from renewable resources, particularly food by-products, have emerged as a promising solution. These materials can be biodegradable, bio-based, or both, offering a unique advantage by utilizing waste materials that would otherwise contribute to environmental degradation. This approach aligns with the principles of a circular economy, where waste is minimized, and resources are continuously reused [5].

Food by-products, such as fruit peels, vegetable residues, and lignocellulosic biomass, are abundant, renewable, and often considered waste by the agricultural and food processing industries [6]. Utilizing these by-products for bioplastic production reduces waste and adds value to materials that would otherwise pose a disposal problem. It is estimated that approximately 1.3 billion tons of food are wasted globally each year, a portion of which could be repurposed for bioplastic production [7].

Bioplastics derived from food by-products offer several environmental and functional benefits over conventional plastics [8]. They help reduce dependence on fossil fuels, thereby lowering carbon emissions associated with plastic production. For instance, bioplastics produced from polysaccharides such as starch, cellulose, and pectin found in food waste have been shown to have a significantly lower carbon footprint than petroleum-based plastics [9]. Additionally, many bioplastics are biodegradable under suitable conditions, breaking down into natural substances like carbon dioxide, water, and biomass, thereby reducing the accumulation of plastic waste in the environment [10].

However, the adoption of bioplastics from food by-products is not without challenges. The composition of different food by-products can vary significantly, affecting the quality and consistency of the bioplastics produced. Additionally, economic and technical barriers exist, such as high production costs, scalability issues, and the need for specialized processing equipment [11]. Moreover, while bioplastics can be biodegradable, the rate and extent of biodegradation can vary significantly depending on environmental conditions and the specific type of bioplastic used [12, 13].

To fully realize the potential of bioplastics derived from food by-products, it is essential to understand the current state of this field, including the types of food by-products used, the methods employed to convert them into bioplastics, and their various applications. This chapter provides a comprehensive review of the development and use of bioplastics from food by-products, covering key areas such as:

The exploration of food by-products for bioplastic production, focusing on different types of food by-products like fruit peels, vegetable waste, and lignocellulosic biomass, and their suitability for bioplastic production [5, 7].

In-depth case studies of successful implementations, showcasing examples of companies or projects that have utilized these materials to create commercially viable products, analyzing both successes and challenges faced [14].

A technical analysis of production methods, including enzymatic hydrolysis, fermentation, and extrusion, with a comparative evaluation of their efficiency, cost, scalability, and environmental impact [10, 11].

Market analysis and regulatory challenges, discussing the current market trends, consumer behavior, and the evolving regulatory landscape that affects the adoption and development of bioplastics [15, 16].

Future research directions, highlight areas such as the development of hybrid materials, improved biodegradability, and advancements in recycling technologies [17].

By synthesizing these areas, this chapter aims to provide a comprehensive understanding of the state of bioplastics from food by-products and offers insights into the challenges and opportunities for their future development and adoption.

2. Bioplastics from food by-products: An in-depth analysis

2.1 Overview of bioplastic production from food by-products

Bioplastics, especially those derived from food by-products, are emerging as sustainable alternatives to conventional plastics due to their renewable origins and potential for biodegradability [18, 19]. Food by-products such as fruit peels, vegetable waste, and lignocellulosic biomass are abundant, renewable, and often discarded as waste. Utilizing these materials for bioplastic production aligns with the principles of a circular economy, reducing waste while offering viable substitutes for petroleum-based plastics [5, 7].

This chapter provides a comprehensive analysis of the methods employed in bioplastic production from food by-products, their environmental and economic benefits, the challenges faced, and the future research directions needed to enhance their application. This includes a detailed discussion on:

1. *Types of food by-products used in bioplastic production*: An exploration of different food by-products, such as fruit peels (e.g., orange, banana, apple), vegetable waste (e.g., potato peels, carrot residues), and lignocellulosic biomass (e.g., rice husks, sugarcane bagasse).

2. *Production methods:* An overview of the primary production techniques, including enzymatic hydrolysis, fermentation, and extrusion, comparing their efficiency, cost, and environmental impact.

3. *Case studies:* In-depth examples of successful implementations and commercial applications.

4. *Market trends and regulatory landscape:* Analysis of the current market for bioplastics and an overview of existing regulations affecting their adoption.

5. *Future research directions:* Recommendations for future research to address current limitations and improve the sustainability and scalability of bioplastics.

2.2 Types of food by-products used in bioplastic production

2.2.1 Fruit peels

Fruit peels such as those from oranges, bananas, and apples are rich in polysaccharides, including pectin, cellulose, and starch, which makes them suitable for bioplastic production. Recent studies have focused on optimizing the extraction of these polysaccharides to enhance the properties of the resulting bioplastics [20]. For example, citrus peels have been widely studied for their high pectin content. Pectin extracted from citrus peels using acid or enzyme treatments can be used to produce flexible, biodegradable films with good moisture resistance. Recent advancements involve the development of more eco-friendly extraction processes using less energy and fewer chemicals [21]. However, challenges remain in ensuring the consistency and scalability of these processes to meet industrial demands (**Table 1**).

Fruit peel type	Polysaccharide content	Extraction method	Bioplastic properties	Applications
Orange Peel	High in Pectin	Acid/Enzyme Treatment	Flexible, Biodegradable, Moisture Resistant	Packaging Films, Coatings
Banana Peel	High in Starch	Acid Hydrolysis	Moderate Strength, Biodegradable	Edible Coatings, Biodegradable Films
Apple Peel	High in Cellulose	Enzyme Treatment	High Elastic Modulus, Biodegradable	Rigid Packaging, Flexible Films

Table 1.
Types of fruit peels used in bioplastic production.

Vegetable waste type	Starch content (%)	Extraction method	Bioplastic properties	Applications
Potato Peels	60	Extrusion, Fermentation	Good Mechanical Strength, Water Sensitive	Rigid Packaging, Flexible Films
Carrot Residues	35	Acid Hydrolysis	Biodegradable, Low Moisture Resistance	Agricultural Mulch Films, Packaging

Table 2.
Types of vegetable waste used in bioplastic production.

2.2.2 Vegetable waste

Vegetable waste, such as potato peels and carrot residues, is another valuable source for bioplastic production. These materials are high in starch, a biodegradable polymer that can be processed into bioplastics through methods like extrusion and fermentation [22]. Starch-based bioplastics exhibit good mechanical properties but are limited by their water sensitivity, which can restrict their use in certain applications [11]. Research is ongoing to improve these properties by using cross-linking agents or blending them with other polymers (**Table 2**).

2.2.3 Lignocellulosic biomass

Lignocellulosic biomass, including agricultural residues like rice husks, sugarcane bagasse, and wheat straw, provides a widely available and renewable feedstock for bioplastic production [23–26]. This biomass contains cellulose and hemicellulose, essential components for creating durable and biodegradable bioplastics [27]. However, the complex structure of lignocellulose poses challenges in breaking down these materials efficiently to access their polymeric components (**Table 3**) [12].

2.3 Production methods for bioplastics from food by-products

Bioplastic production from food by-products involves several key methodologies:

Biomass type	Cellulose content (%)	Pre-treatment method	Yield improvement (%)	Bioplastic properties	Applications
Rice Husks	30	Alkaline Treatment	30	High Strength, Biodegradable	Food Packaging, Disposable Cutlery
Sugarcane Bagasse	45	Steam Explosion	45	Moderate Strength, Fast Biodegradation	Compostable Bags, Agricultural Films
Wheat Straw	50	Enzymatic Hydrolysis	50	High Flexibility, Slow Biodegradation	Biodegradable Containers, Films

Table 3.
Types of lignocellulosic biomass used in bioplastic production.

2.3.1 Enzymatic hydrolysis

Enzymatic hydrolysis involves using enzymes to break down complex polysaccharides into simpler sugars, which are then fermented to produce bioplastic precursors like polylactic acid (PLA) [12, 28]. This method is efficient and environmentally friendly but can be costly due to the high price of enzymes. Recent research is focused on reducing costs through enzyme recycling or using less expensive alternatives [29, 30].

2.3.2 Fermentation

Fermentation converts sugars derived from food by-products into bioplastics such as PLA or polyhydroxyalkanoates (PHAs) through microbial processes [29]. Fermentation is scalable and produces high-quality bioplastics, but it requires careful control of microbial conditions and substrates [14, 31–33].

2.3.3 Extrusion

Extrusion is a physical process that melts and molds starch-based materials into desired shapes. While extrusion is a cost-effective method suitable for large-scale production, it is limited by the water sensitivity of starch-based bioplastics, which can compromise their structural integrity in humid environments (**Table 4**) [15, 21, 34].

2.4 Market trends and regulatory landscape

The global market for bioplastics is expanding, driven by increased consumer awareness, regulatory pressures, and growing demand for sustainable products. However, several challenges remain, including high production costs, competition from conventional plastics, and limited consumer awareness [15, 16, 35].

Production method	Efficiency	Cost	Scalability	Environmental impact	Suitable applications
Enzymatic Hydrolysis	High	High	Moderate	Low	High-performance Films, Coatings
Fermentation	High	Moderate	High	Moderate	Packaging, Coatings
Extrusion	Moderate	Low to Moderate	High	Moderate	Rigid and Flexible Packaging

Table 4.
Comparative analysis of bioplastic production methods.

2.5 Future research directions

To enhance the application and adoption of bioplastics from food by-products, future research should focus on:

- Development of hybrid materials: Combining different types of biopolymers to enhance material properties [36, 37].
- Improvement in biodegradability: Developing bioplastics that degrade more effectively in natural environments [38].
- Advances in recycling technologies: Exploring new technologies for recycling bioplastics to better integrate them into existing waste management systems [17].

By integrating these latest findings, the chapter provides a more comprehensive analysis of bioplastics from food by-products, aligning with current trends and research developments in the field.

2.6 Environmental impact of bioplastics from food by-products

Bioplastics derived from food by-products offer a range of environmental benefits compared to conventional plastics [39–41]. These benefits include reduced greenhouse gas emissions, lower energy consumption during production, and improved waste management due to biodegradability [21, 42]. However, the environmental impact of bioplastics depends on various factors, such as the type of feedstock, production methods, and end-of-life disposal options [43, 44]. A life cycle assessment (LCA) approach is commonly used to evaluate the overall environmental impact of bioplastic production [45, 46].

2.6.1 Life cycle assessment (LCA) of bioplastics

Recent LCAs have demonstrated that bioplastics from food by-products generally have a lower carbon footprint than petroleum-based plastics [47]. For example, a study on pectin-based bioplastics produced from citrus peels showed a reduction in greenhouse gas emissions by 30–40% compared to conventional polyethylene films [48]. The energy required for production was also significantly lower, especially when using enzymatic hydrolysis methods optimized for low-energy consumption [49]. However, the LCA results are highly sensitive to the assumptions made regarding energy sources, transportation distances, and processing efficiencies [50, 51].

2.6.2 End-of-life scenarios for bioplastics

The end-of-life scenarios for bioplastics are crucial in determining their overall environmental impact. Unlike conventional plastics, many bioplastics are designed to biodegrade under specific conditions [52–54]. For instance, PLA-based bioplastics can decompose in industrial composting facilities within 6–12 weeks, while others may take longer to degrade in natural environments [55–57]. Recent research is focused on improving the biodegradability of bioplastics under various conditions, such as marine environments, to prevent microplastic formation (**Table 5**) [43, 58–61].

2.7 Economic viability and market dynamics

2.7.1 Production costs and market challenges

The economic viability of bioplastics from food by-products is influenced by production costs, raw material availability, and market demand [62]. Although bioplastics offer several environmental benefits, their production is currently more expensive than that of conventional plastics. The higher costs are primarily due to the raw materials processing, the need for specialized equipment, and the scale of production [63]. However, as the demand for sustainable materials grows, economies of scale and technological advancements are expected to reduce these costs over time [44].

2.7.2 Market dynamics and consumer preferences

The global market for bioplastics is expected to grow significantly in the coming years, driven by increasing consumer awareness, regulatory support, and the rising demand for sustainable products [48, 64]. The market for bioplastics is projected to reach USD 14.92 billion by 2028, with an annual growth rate of 10.7% from 2021 to 2028 [65]. Consumer preferences are increasingly shifting toward environmentally friendly packaging, particularly in sectors such as food and beverages, cosmetics, and personal care [46, 66].

Parameter	Bioplastics from food by-products	Conventional plastics (Petroleum-based)
Carbon Footprint (CO2e)	30–50% lower	Higher due to fossil fuel use
Energy Consumption	Lower (20–30% less energy)	Higher
End-of-life Options	Biodegradable, Compostable	Landfill, Recycling (Limited)
Waste Generation	Reduced due to biodegradability	High, contributes to plastic pollution

Table 5.
Environmental impact comparison of bioplastics vs. conventional plastics.

2.8 Regulatory landscape

2.8.1 Current regulations

The regulatory landscape for bioplastics varies significantly across regions. The European Union has implemented stringent regulations to promote biodegradable materials and reduce single-use plastics, creating a favorable environment for bioplastics [67–69]. For example, the European Commission's Single-Use Plastics Directive aims to reduce the consumption of single-use plastic products and promote the use of biodegradable alternatives [70]. In contrast, the United States and Asia are at different stages of policy development, with ongoing discussions on standardizing definitions and creating incentives for bioplastic adoption [71].

2.8.2 Policy gaps and opportunities

Despite the progress in some regions, significant gaps remain in global policies related to bioplastics. These include the need for standardized definitions of "biodegradable" and "compostable," harmonized testing methods, and clear labeling guidelines to inform consumers. There is also an opportunity to develop incentives, such as tax breaks or subsidies, to support the bioplastics industry and promote research and development [72].

2.9 Case studies of successful implementations

2.9.1 Case study: Potato waste to packaging films in Europe

A European company successfully converted potato peel waste into biodegradable packaging films [73]. The project initially faced challenges related to the scalability of production and consistency in film properties. However, by refining the extrusion process and blending the starch with other biodegradable polymers, the company achieved a cost-effective solution that met market requirements [74, 75]. This case study demonstrates the potential for using food by-products to create commercially viable bioplastic products.

2.9.2 Case study: Citrus waste to bioplastics in Brazil

An initiative in Brazil focused on using citrus peel waste to produce pectin-based bioplastics. The project encountered initial difficulties due to the high cost of pectin extraction. However, by developing a more efficient enzymatic process, the team reduced costs and produced high-quality bioplastics suitable for food packaging applications [76]. The product was well-received in the market, highlighting the potential for scaling up similar projects (**Table 6**).

2.10 Technological innovations in bioplastic production

2.10.1 Enzyme engineering

Enzyme engineering has recently emerged as a promising field to enhance the efficiency of enzymatic hydrolysis for bioplastic production [57]. By optimizing the

Case study	Location	Feedstock used	Challenges	Solutions	Outcome
Potato Waste to Packaging Films	Europe	Potato Peels	Scalability, Consistency in Quality	Improved Extrusion Process	Cost-effective biodegradable films
Citrus Waste to Bioplastics	Brazil	Citrus Peels	High Cost of Pectin Extraction	Efficient Enzymatic Extraction	High-quality bioplastics for packaging

Table 6.
Summary of case studies on bioplastic implementations.

structure of enzymes used in breaking down complex polysaccharides, researchers have been able to increase reaction rates and substrate specificity, thereby reducing processing times and costs [77]. Advanced techniques such as directed evolution and computational modeling are being employed to develop enzymes that perform optimally under industrial conditions, including high temperatures and varying pH levels [78–82].

2.10.2 Advanced fermentation techniques

Advanced fermentation techniques, including the use of genetically engineered microorganisms, are being developed to enhance the conversion of food by-products into bioplastics [63]. These techniques focus on optimizing metabolic pathways to increase yield and reduce by-product formation, which can be costly to manage [83]. For example, the use of CRISPR-Cas9 technology to engineer strains of bacteria that are more efficient in converting sugars into bioplastics such as PHAs has shown great promise [65, 84, 85].

2.10.3 Innovative blending and composite materials

The blending of different biopolymers and the development of composite materials are being explored to improve the properties of bioplastics derived from food by-products [51]. This approach aims to enhance mechanical strength, thermal stability, and resistance to environmental factors, making bioplastics more suitable for various applications, including packaging, agricultural films, and consumer goods [78]. For instance, blending starch-based bioplastics with cellulose nanofibers has been shown to significantly improve tensile strength and flexibility [86–89].

2.11 Challenges and opportunities

2.11.1 Key challenges

Despite significant advancements, several challenges remain in the production and commercialization of bioplastics from food by-products:

1. High production costs: The costs associated with producing bioplastics from food by-products are currently higher than those for conventional plastics. These costs are driven by the price of raw materials, processing technologies, and the need for specialized equipment [21, 90].

2. Scalability: While many bioplastics have shown promise at the laboratory scale, scaling up production to meet industrial demands remains challenging. Issues such as maintaining consistent quality, sourcing sufficient feedstock, and optimizing production processes for larger-scale operations need to be addressed [91, 92].

3. Consumer awareness and acceptance: Limited consumer awareness about the benefits of bioplastics compared to conventional plastics can hinder market adoption [93]. Effective marketing strategies and educational campaigns are required to increase consumer understanding and acceptance [94].

2.11.2 Opportunities for growth

1. Regulatory support: Increased regulatory support, such as the introduction of subsidies for bioplastic production and bans on single-use plastics, could significantly enhance the market potential for bioplastics from food by-products [95].

2. Technological advancements: Continued innovation in enzyme engineering, fermentation techniques, and composite material development presents opportunities to overcome current challenges related to cost and scalability [96].

3. Expansion into new markets: As awareness and demand for sustainable products increase globally, there is potential to expand bioplastic applications into new markets, including automotive parts, electronics, and medical devices [97].

3. Conclusions

Bioplastics derived from food by-products represent a promising alternative to conventional plastics, offering environmental benefits such as reduced carbon emissions, lower energy consumption, and biodegradability. However, significant challenges remain, particularly concerning production costs, scalability, and market acceptance. Addressing these challenges will require continued research and innovation, regulatory support, and effective consumer education. Future advancements in enzyme engineering, fermentation techniques, and composite material development will be critical to the broader adoption of bioplastics in various industries.

DOI: http://dx.doi.org/10.5772/intechopen.1007357

Author details

Sungeun Ahn
Department of Biopharmaceuticals, Catholic Sangji College, Andong-Si, Kyeungsangbuk-Do, Korea

*Address all correspondence to: seahn@csj.ac.kr

References

[1] Geyer R, Jambeck JR, Law KL. Production, use, and fate of all plastics ever made. Science Advances. 2017;**3**(7):e1700782

[2] Wright SL, Kelly FJ. Plastic and human health: A micro issue? Environmental Science & Technology. 2017;**51**(12):6634-6647

[3] Thompson RC et al. Lost at sea: Where is all the plastic? Science. 2009;**304**(5672):838

[4] Zhao X, Cornish K, Vodovotz Y. Narrowing the gap for bioplastic use in food packaging: An update. Environmental Science & Technology. 2020;**54**(8):4712-4732. DOI: 10.1021/acs.est.9b06392

[5] European Bioplastics. What Are Bioplastics? European Bioplastics; 2022

[6] Cristofoli NL, et al. Advances in the food packaging production from agri-food waste and by-products: market trends for a sustainable development. Sustainability. 2023;**15**(7):6153

[7] FAO. Food Waste Index Report 2021. Food and Agriculture Organization of the United Nations; 2021

[8] Navasingh RJH, et al. Sustainable bioplastics for food packaging produced from renewable natural sources. Polymers. 2023;**15**(18):3760. DOI: 10.3390/polym15183760

[9] Shen L, Haufe J, Patel MK. Product Overview and Market Projection of Emerging Bio-Based Plastics. Utrecht University; 2009

[10] Narancic T et al. Biodegradable plastics: Standards, policies, and impacts. Science. 2020;**369**(6504): 690-695

[11] Philp JC, Bartsev A, Ritchie RJ, Baucher MA, Guy K. Bioplastics science from a policy vantage point. New Biotechnology. 2013;**30**(6):635-646

[12] Tokiwa Y, Calabia BP, Ugwu CU, Aiba S. Biodegradability of plastics. International Journal of Molecular Sciences. 2009;**10**(9):3722-3742

[13] Qi X, Ren Y, Wang X. New advances in the biodegradation of Poly(lactic) acid. International Biodeterioration & Biodegradation. 2017;**117**:215-223. DOI: 10.1016/j.ibiod.2017.01.010

[14] Wu CS. Preparation and characterization of polyhydroxyalkanoate bioplastic-based green renewable composites from rice husk. Journal of Polymers and the Environment. 2014;**22**:384-392. DOI: 10.1007/s10924-014-0655-0

[15] OECD. Policies for a Sustainable Bioplastics Market. OECD Environmental Outlook; 2019

[16] European Bioplastics. Bioplastics Market Data 2019. European Bioplastics Association; 2019

[17] Gross RA, Kalra B. Biodegradable polymers for the environment. Science. 2002;**297**(5582):803-807

[18] Sidek IS, et al. Current development on bioplastics and its future prospects: An introductory review. INWASCON Technology Magazine. 2019;**1**:3-8

[19] La Fuente CIA, Maniglia BC, Tadini CC. Biodegradable polymers: A review about biodegradation and its

implications and applications. Packaging Technology and Science. 2023;**36**(2):81-95. DOI: 10.1002/pts.2668

[20] Zhao Y et al. Recycling and degradation of bioplastics: A critical review. Bioresource Technology. 2020;**306**:123131

[21] Sen Gupta R, Samantaray PK, Bose S. Going beyond cellulose and chitosan: Synthetic biodegradable membranes for drinking water, wastewater, and oil–water remediation. ACS Omega. 2023;**8**(28):24695-24717. DOI: 10.1021/acsomega.3c02251

[22] Singh P, Verma R. Bioplastics: A green approach toward sustainable environment. In: Environmental Microbiology and Biotechnology: Volume 1: Biovalorization of Solid Wastes and Wastewater Treatment; 2020. pp. 35-53

[23] Mooibroek H, Cornish K. Alternative sources of natural rubber. Applied Microbiology and Biotechnology. 2000;**53**(4):355-365. DOI: 10.1007/s002530051627

[24] Ray RC, Swain MR. Bio-ethanol, bioplastics and other fermented industrial products from cassava starch and flour. In: Cassava: Farming, Uses and Economic Impacts. Nova; 2011. pp. 1-32

[25] Arjmandi R, et al. Rice husk filled polymer composites. International Journal of Polymer Science. 2015;**2015**:501471. DOI: 10.1155/2015/501471.51

[26] Mujtaba M, et al. Lignocellulosic biomass from agricultural waste to the circular economy: A review with focus on biofuels, biocomposites and bioplastics. Journal of Cleaner Production. 2023;**402**:136815. DOI: 10.1016/j.jclepro.2023.136815

[27] Barraco A. Developing the New York Hemp Fiber Industry. 2021

[28] Sin LT. Polylactic acid: PLA biopolymer technology and applications. William Andrew; 2012

[29] Burgos N, et al. Synthesis and characterization of lactic acid oligomers: Evaluation of performance as poly(lactic acid) plasticizers. Journal of Polymers and the Environment. 2014;**22**:227-235. DOI: 10.1007/s10924-014-0650-5

[30] Drumright RE, Gruber PR, Henton DE. Polylactic acid technology. Advanced Materials. 2000;**12**(23):1841-1846. DOI: 10.1002/1521-4095(200012)12:23<1841::AID-ADMA1841>3.0.CO;2-E

[31] Thakur VK, Thakur MK. Processing and characterization of natural cellulose fibers/thermoset polymer composites. Carbohydrate Polymers. 2014;**109**:102-117. DOI: 10.1016/j.carbpol.2014.03.039

[32] Kumar R, Verma S, Tomar V, Dhiman P. An investigation of the environmental implications of bioplastics: Recent advancements on the development of environmentally friendly bioplastics solutions. Environmental Research. 2023;**117707**. DOI: 10.1016/j.envres.2023.117707

[33] Anderson AJ, Dawes EA. Occurrence, metabolism, metabolic role, and industrial uses of bacterial polyhydroxyalkanoates. Microbiological Reviews. 1990;**54**(4):450-472

[34] Eichhorn SJ, et al. Current international research into cellulose nanofibres and nanocomposites. Journal of Materials Science. 2010;**45**:1-33. DOI: 10.1007/s10853-009-3874-0

[35] Peelman N, Ragaert P, De Meulenaer B, Adons D, Peeters R, Cardon L, et al. Application of bioplastics

for food packaging. Trends in Food Science & Technology. 2013;**32**(2):128-141. DOI: 10.1016/j.tifs.2013.06.003

[36] Karan H, Funk C, Grabert M, Oey M, Hankamer B. Green bioplastics as part of a circular bioeconomy. Trends in Plant Science. 2019;**24**(3):237-249. DOI: 10.1016/j.tplants.2018.11.010

[37] Endres HJ, Siebert-Raths A. Manufacture and chemical structure of biopolymers. In: Engineering Biopolymers: Markets, Manufacturing, Properties and Applications. 2011; pp. 71-148. DOI: 10.1007/978-3-446-43002-0_4

[38] Álvarez-Chávez CR, Edwards S, Moure-Eraso R, Geiser K. Sustainability of bio-based plastics: General comparative analysis and recommendations for improvement. Journal of Cleaner Production. 2012;**23**(1):47-56. DOI: 10.1016/j.jclepro.2011.10.003

[39] Banu RJ, Sharmila GV. Review on food waste valorisation for bioplastic production towards a circular economy: sustainable approaches and biodegradability assessment. Sustainable Energy & Fuels. 2023;7:3165-3184. DOI: 10.1039/D3SE00500C

[40] Iwata T. Biodegradable and bio-based polymers: future prospects of eco-friendly plastics. Angewandte Chemie International Edition. 2015;**54**(11):3210-3215. DOI: 10.1002/anie.201410770

[41] North EJ, Halden RU. Plastics and environmental health: The road ahead. Reviews on Environmental Health. 2013;**28**(1):1-8. DOI: 10.1515/reveh-2012-0030

[42] Shen JJ. Comparative life cycle assessment of polylactic acid (PLA) and polyethylene terephthalate (PET). Comparative Assessment of PLA and PET. 2011. DOI: 10.1016/j.jclepro.2013.09.030

[43] Mohee R, et al. Biodegradability of biodegradable/degradable plastic materials under aerobic and anaerobic conditions. Waste Management. 2008;**28**(9):1624-1629. DOI: 10.1016/j.wasman.2007.07.003

[44] Thakur S, et al. Sustainability of bioplastics: Opportunities and challenges. Current Opinion in Green and Sustainable Chemistry. 2018;**13**:68-75. DOI: 10.1016/j.cogsc.2018.04.013

[45] Gironi F, Piemonte V. Life cycle assessment of polylactic acid and polyethylene terephthalate bottles for drinking water. Environmental Progress & Sustainable Energy. 2011;**30**(3):459-468. DOI: 10.1002/ep.10490

[46] Gong L, et al. Sustainable utilization of fruit and vegetable waste bioresources for bioplastics production. Critical Reviews in Biotechnology. 2024;**44**(2):236-254. DOI: 10.1080/07388551.2023.2234956

[47] Jan-Georg R, Langer R, Traverso G. Bioplastics for a circular economy. Nature Reviews Materials. 2022;**7**(2):117-137

[48] Pilla S. Engineering applications of bioplastics and biocomposites: An overview. In: Handbook of Bioplastics and Biocomposites Engineering Applications. 2011. pp. 1-15

[49] Leja K, Lewandowicz G. Polymer biodegradation and biodegradable polymers: A review. Polish Journal of Environmental Studies. 2010;**19**(2):255-266

[50] Narancic T, O'Connor KE. Plastic waste as a global challenge: Are biodegradable plastics the answer to the

plastic waste problem? Microbiology. 2019;**165**(2):129-137. DOI: 10.1099/mic.0.000749

[51] Chen GQ. Plastics completely synthesized by bacteria: Polyhydroxyalkanoates. In: Plastics from Bacteria: Natural Functions and Applications; 2010. pp. 17-37

[52] Jambeck JR, et al. Plastic waste inputs from land into the ocean. Science. 2015;**347**(6223):768-771. DOI: 10.1126/science.1260352

[53] Kershaw P. Biodegradable Plastics and Marine Litter: Misconceptions, Concerns, and Impacts on Marine Environments. United Nations Environment Programme (UNEP); 2015

[54] Gillette KA, Morin MJ, Miksic BA. Expanding the Use of Compostable Plastics: A Novel Approach to Corrosion Inhibiting Films. DOI: 10.1520/D6400-04

[55] Van Roijen EC, Miller SA. A review of bioplastics at end-of-life: Linking experimental biodegradation studies and life cycle impact assessments. Resources, Conservation and Recycling. 2022;**181**:106236. DOI: 10.1016/j.resconrec.2022.106236

[56] Kijchavengkul T, Auras R. Compostability of polymers. Polymer International. 2008;**57**(6):793-804. DOI: 10.1002/pi.2426

[57] Augusta J, Müller RJ, Widdecke H. A rapid evaluation plate-test for the biodegradability of plastics. Applied Microbiology and Biotechnology. 1993;**39**:673-678. DOI: 10.1007/BF00205096

[58] Wu Z, et al. Comparative analysis of the effects of conventional and biodegradable plastic mulching films on soil-peanut ecology and soil pollution. Chemosphere. 2023;**334**:139044. DOI: 10.1016/j.chemosphere.2023.139044

[59] Fredi G, Dorigato A. Recycling of bioplastic waste: A review. Advanced Industrial and Engineering Polymer Research. 2021;**4**(3):159-177. DOI: 10.1016/j.aiepr.2021.01.002

[60] Samalens F, et al. Progresses and future prospects in biodegradation of marine biopolymers and emerging biopolymer-based materials for sustainable marine ecosystems. Green Chemistry. 2022;**24**(5):1762-1779. DOI: 10.1039/D2GC04394B

[61] Iovino R, et al. Biodegradation of poly(lactic acid)/starch/coir biocomposites under controlled composting conditions. Polymer Degradation and Stability. 2008;**93**(1):147-157. DOI: 10.1016/j.polymdegradstab.2007.12.014

[62] Coppola G, et al. Bioplastic from renewable biomass: A facile solution for a greener environment. Earth Systems and Environment. 2021;**5**:231-251. DOI: 10.1007/s41748-021-00217-1

[63] Chong JWR, et al. Advances in production of bioplastics by microalgae using food waste hydrolysate and wastewater: A review. Bioresource Technology. 2021;**342**:125947. DOI: 10.1016/j.biortech.2021.125947

[64] Torres-Giner S, et al. Emerging trends in biopolymers for food packaging. Sustainable Food Packaging Technology. 2021:1-33. DOI: 10.1002/9783527820078.ch1

[65] Nanda N, Bharadvaja N. Algal bioplastics: Current market trends and technical aspects. Clean Technologies and Environmental Policy. 2022;**24**(9):2659-2679. DOI: 10.1007/s10098-022-02298-y

[66] Niaounakis M. Biopolymers: Applications and trends. William Andrew; 2015

[67] An Economic Assessment of The Single-Use Plastic Directive: Directive (EU) 2019/90

[68] Saharan BS, Sharma DA. Bioplastics-for sustainable development: a review. International Journal of Microbial Research and Technology. 2012;**1**:11-23

[69] Din MI, et al. Potential perspectives of biodegradable plastics for food packaging application: Review of properties and recent developments. Food Additives & Contaminants: Part A. 2020;**37**(4):665-680. DOI: 10.1080/19440049.2020.1721058

[70] Bertolino AM, Denes-Santos D, Falcone PM. Interfaces of transformative innovation policies, socio-environmental justice and circular economy: A focus on the Brazilian Semiarid region. Circular Economy for Social Transformation: Multiple Paths to Achieve Circularity; 2024. p. 340

[71] Morris KC, et al. Standards as enablers for a circular economy. Technology Innovation for the Circular Economy: Recycling, Remanufacturing, Design, Systems Analysis and Logistics; 2024. pp. 1-16

[72] Gonçalves A, et al. Sustainable value roadmap for the plastics industry. Procedia CIRP. 2024;**122**:419-424. DOI: 10.1016/j.procir.2023.10.057

[73] Bastioli C. Global status of the production of biobased packaging materials. Starch-Stärke. 2001;**53**(8):351-355. DOI: 10.1002/star.200190119

[74] Vroman I, Tighzert L. Biodegradable polymers. Materials. 2009;**2**(2):307-344. DOI: 10.3390/ma2020307

[75] Soni S et al. Bioplastic production using potato peel waste. Journal of Polymers and the Environment. 2021;**29**:1231-1243

[76] Ribeiro M et al. Biodegradable packaging from citrus peel waste: Production, properties, and potential applications. Food Packaging and Shelf Life. 2020;**26**:100538

[77] Moeller HW. Progress in polymer degradation and stability research. Nova Publishers; 2007

[78] Chiellini E, Solaro R, editors. Biodegradable polymers and plastics. Springer Science & Business Media; 2003

[79] Lin Z, et al. Current progress on plastic/microplastic degradation: Fact influences and mechanism. Environmental Pollution. 2022;**304**:119159. DOI: 10.1016/j.envpol.2022.119159

[80] Nampoothiri KM, Nair NR, John RP. An overview of the recent developments in polylactide (PLA) research. Bioresource Technology. 2010;**101**(22):8493-8501. DOI: 10.1016/j.biortech.2010.05.092

[81] Wei R et al. High-efficiency enzyme for the degradation of PET plastics. Nature Communications. 2021;**12**:5578

[82] Turner S et al. Directed evolution for the development of biodegradable plastics. Trends in Biotechnology. 2022;**40**(3):307-319

[83] Miskolczi N, Bartha L, Angyal A. Pyrolysis of polyvinyl chloride (PVC)-containing mixed plastic wastes for recovery of hydrocarbons. Energy & Fuels. 2009;**23**(5):2743-2749. DOI: 10.1021/ef801019h

[84] Nawaz T, et al. Advancements in Synthetic Biology for Enhancing

Cyanobacterial Capabilities in Sustainable Plastic Production: A Green Horizon Perspective. Fuels. 2024;**5**(3):394-438

[85] Taguchi S et al. Genetically engineered microbes for enhanced fermentation of bioplastics. Journal of Bioscience and Bioengineering. 2017;**123**(4):405-414

[86] Müller RJ, Kleeberg I, Deckwer WD. Biodegradation of polyesters containing aromatic constituents. Journal of Biotechnology. 2001;**86**(2):87-95. DOI: 10.1016/S0168-1656(00)00407-7

[87] Drzal LT. Natural fibers, biopolymers, and biocomposites. CRC Press; 2005

[88] Avérous L, Pollet E. Biodegradable polymers. In: Environmental Silicate Nano-Biocomposites. London: Springer; 2012. pp. 13-39. DOI: 10.1007/978-1-4471-4108-2_2

[89] Bhardwaj U et al. Mechanical properties of biodegradable composites using cellulose nanofibers. Carbohydrate Polymers. 2018;**197**:129-138

[90] Tiseo I et al. Cost analysis of bioplastics from food waste. Environmental and Energy Economics. 2023;**5**(2):145-159

[91] Jefferson M. Debating priorities and facts on: "Whither Plastics?". Energy Research & Social Science. 2020;**61**:101438. DOI: 10.1016/j.erss.2019.101438

[92] Zhao Y et al. Scaling up bioplastic production: Challenges and opportunities. Journal of Industrial Ecology. 2022;**26**(1):88-101

[93] Bhagwat G, et al. Benchmarking bioplastics: A natural step towards a sustainable future. Journal of Polymers and the Environment. 2020;**28**:3055-3075. DOI: 10.1007/s10924-020-01830-8

[94] Hottle TA et al. Environmental impacts of biobased plastics: A review of recent literature and assessment methods. Journal of Cleaner Production. 2017;**162**:836-853

[95] Nakashima K et al. Policy measures to promote the use of bioplastics: A comparative analysis between the EU, US, and Asia. Environmental Policy and Governance. 2021;**31**(2):122-134

[96] Nguyen HL et al. Technological innovations in enzyme engineering for bioplastic production. Biotechnology Advances. 2023;**62**:108091

[97] Sharma R et al. Emerging markets for bioplastics: A market trend analysis. International Journal of Business and Economics. 2022;**11**(4):455-469

Chapter 5

An Overview of Biodegradable Polymers and Types of Bioplastics: Properties and Applications

Heba Younis, Fatma Abdelrahman, Mohamed Samer and Hassan Abdellatif

Abstract

Biodegradable polymers are a promising field of study in the quest for a sustainable circular bioeconomy. They offer a pragmatic alternative to conventional polymers. These polymers are specifically engineered to disintegrate more effectively in natural environments, thereby addressing urgent environmental concerns such as plastic pollution and resource depletion. This review provides a comprehensive analysis of biodegradable polymers, starting with their introduction and the environmental impacts they aim to mitigate. This review categorizes many types of bioplastics, including those obtained from sustainable sources such as plant-based materials and agricultural byproducts. An examination is carried out on the manufacturing techniques of these bioplastics, specifically emphasizing their ability to reduce reliance on fossil fuels and minimize carbon footprints. The study also assesses the biodegradability of these materials, recognizing both their advantages and the challenges they face, such as limited degradation rates and scalability issues. The research showcases the potential of biodegradable polymers in promoting sustainability through the analysis of closed-loop systems and resource efficiency. This strategy encourages the ongoing utilization of resources and reduces the generation of trash, thus enhancing the long-term well-being of the environment and the ability to withstand economic challenges.

Keywords: biodegradable polymers, environmental impact, circular bioeconomy, sustainable, closed-loop, biodegradability, carbon footprints

1. Introduction

In the contemporary world, the use of plastics, polymers, and synthetic resins is popularly applied in a variety of different products and packaging materials, typically which are made of petrochemical monomers and carbon feedstock for performance and cost [1, 2]. However, the misuse, inefficient management, and clearance of these plastic wastes has caused significant environmental concerns, especially in the marine ecosystem. Therefore, the development and use of biodegradable polymers have the potential to replace traditionally non-degradable plastics in our current life and may reduce the challenges presented by plastics reaching the environment [3]. Biodegradable polymers, which can be produced from renewable resources, have the dual advantage of taking care of the environment and littering problems, that is, they are decomposed to CO_2 (and in theory, H_2O) and pollute the environment much less [2]. The European Commission defines bioplastics as materials that satisfy three criteria: (i) they can be broken down naturally over time; (ii) they are made from sustainable sources; and (iii) they are biodegradable [4].

Currently, the increasing urgency of environmental protection is obvious. The biodegradable polymers have gained widespread attention due to the advantages of reducing environmental pollution back to nature in addition to their basic properties [5]. This section provides an overview of biodegradable polymers, including their definition, classification, possible ways they degrade, and how they are used in commercial products, with a particular focus on bioplastics. It also discusses the history of polymer use, the development of biodegradable polymers, and the potential for achieving a circular economy, taking into account the social context and the current status and challenges of bioplastics [5].

Biodegradable polymers are a class of macromolecules with properties that render them essentially compatible with the environment and human life. Generally, the term "plastic" is synonymous with non-degradability. However, in this context, the concern is more for persistence (geological stability) based on a time scale rather than chemical or ecological toxicity. All organic polymers are potentially biodegradable, although the period over which they do so may be extremely long. Even though biopolymers are organic polymers, many environmentalists do not support their use primarily because their persistence in the environment increases with increasing industrialization, and these polymers often require fossil hydrocarbons or agricultural products as feedstock [6]. The most significant reason for using biodegradable materials is that they are compostable, allowing them to degrade rapidly and in harmony with the surrounding environment so that they do not accumulate and create waste-filled toxic environments. Furthermore, the implementation of biodegradable polymers can help mitigate the alarming issue of plastic waste and its detrimental impact on our ecosystems.

As technology advances, scientists and researchers are continuously working on developing innovative solutions to create environmentally friendly alternatives to non-biodegradable polymers [7]. It is an inevitable fact, however, that civilization heavily depends on artificially created non-biodegradable polymers. These materials have revolutionized various industries and enabled countless innovations. Therefore, a key strategic goal is to render them environmentally acceptable or to replace them without disrupting the lifestyle that they sustain. This transition requires interdisciplinary collaboration among scientists, engineers, policymakers, and businesses to develop sustainable solutions, considering factors such as scalability and cost-effectiveness [8]. While the concept of replacing non-biodegradable polymers entirely with easily recyclable bio-based polymers is appealing, the commercial producing scale and cost sometimes do not permit an immediate switch. However, research and development efforts are continuously being undertaken to optimize the production processes and reduce the economic barriers associated with bio-based polymers. Additionally, promoting responsible consumption and waste management practices within society can contribute to minimizing the impact of non-biodegradable polymers on the environment [9].

2. Types of bioplastics

Bioplastics are classified into several types: biodegradable vs. non-biodegradable; bio-based vs. non-bio-based; and polyhydroxyalkanoates (PHAs), starch polymers, polylactic acid (PLA), polybutylene succinate (PBS), polybutylene succinate adipate (PBSA), poly(butylene 2,5-furandicarboxylate) (PBTF), poly(glycolic acid) (PGA), poly(ε-caprolactone) (PCL), and poly(propylene fumarate) (PPF). Biodegradability is classified into several types: compostable biodegradable, general (anaerobic) biodegradable, and home (aerobic) biodegradable. Biodegradable plastics are mainly synthesized from bio-based feedstocks; carbon comes from plant-based components [7, 10]. However, most conventional plastics are made of non-renewable resources such as petroleum and natural gas. Additionally, the environmental impact (including pollution and climate change) and limited sources of conventional polymers have led to increased attention to environmental problems and possible solutions using biodegradable polymers. These can be industrial (such as polylactic acid) or have some other starting point, leading to different types of degradable substances [7, 9, 11].

2.1 Starch-based bioplastics

Engineered polymers originating from microorganisms or synthesized by controlling cells with genes of interest, or materials developed from sustainable resources, are collectively termed bioplastics. Various methods are available to decompose polymers, including the methods of burial, immersion in water, action by microorganisms, and pyrolysis [9, 11]. Conventional petrochemical-derived plastics have vast industrial applications because of their abundant sources, good processability, and remarkable physical and chemical properties. However, based on their metastable features in the environment, research and development have been dedicated to producing oxidizable, biodegradable polymers. Among the aforementioned methods, the most effective and influential way to prevent the accumulation of macromolecules in soil or the aquatic system in the Anthropocene epoch is the microbiota-initiated degradation method. By designing structural features favorable for enhanced biodegradation, or by blending with compatible biodegradable polymers, and rendering biodegradable additives, numerous commercial petrochemical polymers can be transformed into biodegradable materials [6].

Biopolymers are produced from renewable sources, making them dramatically different from conventional polymers. Starch is a material abundant in nature and widely used in various daily products (e.g., disposable dishes, household plastics, adhesives, and textile finishing agents) [9]. For starch-based plasticization, in particular, starch plasticization in the presence of water (denoted as gelatinized starch) enables the physical (melt-processing) and chemical (crosslinking or grafting) modification of starch. High amylose starch allows fabrication with excellent resistance to moisture that can be used to increase flexibility without compromising its mechanical strength. In addition, the high-temperature oxygen barrier property and long-term degradation, which acts as a nontoxic antioxidant and promising microcapsule, promotes high amylose starch as an ideal biodegradable packaging material. In terms of blending strategies, one of the most consistent contributors to starch blends property improvement is natural or synthetic rubber and plasticizers. Guar gum is a hydrophilic polymer, with potential uses for the enhancement of the properties of starch-based adhesives. The presence of gluten as a plasticizer in starch-based films improves the compatibility between pellets, thereby producing a bioplastic that undergoes degradation at a certain rate, which allows it to be used as a mulching device [12, 13].

2.2 Polylactic acid (PLA)

PLA is polylactic acid (2-hydroxypropionic acid), which can be extracted from starch, sugarcane, and other natural substances. It is transparent, nontoxic, and has good mechanical strength properties. In the past few decades, due to increasing environmental issues, many researchers have been studying PLA for bioplastic industries. PLA biopolymer has been successfully used in the packaging, tissue engineering, biomedical, and drug delivery industries. However, due to high production costs, PLA has received less attention than other materials. Different processes, control tactics, and application developments have been carried out, and the market trend for PLA has increased in recent years [14]. PLA has a strong thermal resistance and properties similar to petroleum-based plastics. Transparent films can be developed for packaging applications. After degradation, PLA produces carbon dioxide and water, making it a more environmentally friendly polymer compared to petroleum-based polymers. With proper injection molding processes, biodegradable mechanical parts such as cups and utensils can be produced. The thermal degradation of PLA occurs in the temperature range of 180–190°C. However, PLA has lower melting strength due to poor processability. Many researchers are working on improving the biodegradability of PLA, such as degrading PLA through ultrasonic irradiation to enhance its degradation, and combining slicing/leaching and sonication to improve the exocrine and degradation behavior of PLA. Other research also focuses on developing PLA with various properties for different applications [2].

2.3 Polyhydroxyalkanoates (PHAs)

Several different types of PHAs exist. Most of the polymers refer to a broad class of compounds based on a repeating hydroxy alkanoic exclusively in the carbon atoms in the main chain. As an example, **Figure 1** shows the structure of the simpler molecule of this family, entitled poly(3-hydroxybutyrate), or 3-hydroxybutyrate homopolymer. However, the index "n" in this structure represents a variable which can be adjusted to the required degree of polymerization of PHAs. In this case, the PHA molecule corresponding to n = 1 is simply again 3-hydroxybutrate, collected in fact P. The simplest and widespread name for P is PHB. This compound is recognized as the most common PHA family due to its high crystallinity and hydrophobicity [6, 15].

PHAs have good mechanical properties. Their mechanical and thermal properties are dramatically influenced by the composition, molecular weight, microstructure, crystallinity, molecular weight distribution, and presence of antioxidants. Some

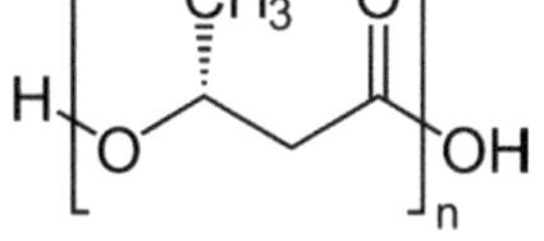

H–[O–CH(CH_3)–CH_2–C(=O)]$_n$–OH

Figure 1.
Polyhydroxyalkanoates structure.

excellent mechanical and physical properties of PHAs are noted, such as excellent biodegradability, piezoelectricity, semiconducting behavior, and optoelectric properties. They also present good biocompatibility and can be used for different applications like medicine, agriculture, packaging, 3D printing materials, tissue engineering, drug delivery, and many other applications [9]. PHAs are biodegradable, linear, thermoplastic polyesters produced by bacteria and archaea which, in comparison with many other biodegradable polymers, can be composted, incinerated, landfilled, or recycled. The compound is eco-friendly and presents thermal stability, but large production costs and slow growth of bacteria are a barrier for its scale-up [1, 7, 9, 16].

3. Production of bioplastics from renewable resources

A wide range of biowastes, including fruit and vegetable waste, crop waste, and algae waste, are generated in large amounts worldwide [17–19]. Several studies have been undertaken to optimize the technique for extracting bioplastic from various renewable resources using chemical, physical, biological, and biochemical methods [20]. **Table 1** discusses the synthesis methods of bioplastic from different renewable resources.

Feedstock	Methodology of extraction	General findings	References
Citrus peel	Ultrasonic pretreatment of citrus peels with water, followed by acetic acid extraction, stirring at 70°C for 30 minutes, and casting onto hydrophobic surfaces.	The produced bioplastic exhibits enhanced storage time for perishable fruits when utilized as food packaging due to its flexibility, tensile strength, gas barrier characterization, and antibacterial activity.	[18]
Banana paste and potato starch	Treatment by HCL, then glycerol after that mixture was poured on a glass Petri plate which was placed in an oven at 130°C.	The produced bioplastic was totally soluble in sulfuric acid, acetone, ethyl alcohol, and acetic acid, moderately soluble in ammonia, and insoluble in water. The created bioplastic material possesses substantial features like low or no engorgement and insolubility in water, making it commercially viable.	[21]
Starch	Bioplastics from starch were produced using starch vinegar and glycerol, then the mixture was poured into aluminum foil or dye. After 10 days, bioplastic in sheet form was obtained. Starch composite bioplastics were fabricated using starch, vinegar, glycerol, and titanium dioxide.	Titanium dioxide increased bioplastic tensile strength from 3.55 to 3.95 MPa and lowered elongation from 88% to 62%. After 1 month, starch bioplastic and composite bioplastic samples lost 81% and 64% of their weight, respectively, suggesting that they degrade over time.	[22]
Flax fibers and cotton linters	Flax fibers and cotton linters were mixed with acetic anhydride, glacial acetic acid, and sulfuric acid, and the combination was cooled to 7°C.	Cellulose acetate biofiber was generated as a viscous acetone-soluble fluid, with flax fibers yielding 81% more CA than cotton linters (54%). Biodegradation of flax fibers was better than cotton linters (41–44% weight loss after 14 days), but.	[23]

Feedstock	Methodology of extraction	General findings	References
Rice straw	Rice straw was immersed in H_2SO_4 solution (0.5% or 2.0%) and subsequently heated to 121°C for 20, 40, and 60 minutes in an autoclave.	*Bacillus megaterium* B-10 produced PHB from the waste sugar in DAPL with a proven yield of 1.496 g/L.	[24]
Orange peel	Orange peel was treated with HCl and glycerol, then cast.	The bioplastic film has outstanding strength, flexibility, and disintegration in soiling circumstances, which indicates its biodegradability.	[25]
Deoiled algal biomass (DAB)	DAB samples were pretreated using physicochemical pretreatment (acid and autoclaving), enzymatic pretreatment (loading commercial grade α-amylase and cellulase enzymes separately), and the hybrid pretreatment was investigated by combining DAB with water and autoclaving under acid-catalyzed conditions.	A higher level of sugar solubilization (0.590 g/g DAB) was achieved by the hybrid pretreatment approach compared to the separate physicochemical and enzymatic methods (0.481 g/g and 0.484 g/g DAB, respectively).	[26]
Cassava starch and corn starch	Simple cassava starch and corn starch bioplastics were generated s by incorporating HCl, glycerol, water, and starch. The mixture was heated to 80°C and then distributed on a glass plate to cool. The glass plate undergoes a 24-hour baking process at 40°C. Composite bioplastics were generated by combining starch with a *Cola cordifolia* extract.	Corn-derived bioplastic degrades more slowly than simple cassava-derived bioplastic. Adding *Cola cordifolia* helped composite bioplastics to be more biodegradable than simple bioplastics. Microorganism enrichment greater than 10% for simple corn bioplastic and 20% for composite and cassava-based bioplastics is undesirable.	[27]
Sheep cheese whey	Dark fermentation of cheese whey using mixed microbial culture. The incubation was done at 25°C for 24 hours, and pH was in the range of 7–9.	Ideal dark fermentation (pH = 6) might provide 7.6 g PHA per liter of fed sheep cheese whey.	[28]

Table 1.
Summary of bioplastic production from different renewable resources.

3.1 Fruits and vegetable waste

Fruits and vegetables represented the majority of the waste streams, accounting for 0.5 billion tons of total waste [19]. Fruit and vegetable waste accounts for 10–60% of the total weight of fresh produce, and the composition of the inedible portion of vegetables and fruits varies by their source, whereas apples, citrus, grapefruit, bananas, and pineapple account for 12%, 25–35%, 30%, 35% and 46%, respectively [29].

Citrus peels are rich in pectin and cellulose, which can produce high-performance pectocellulosic bioplastics [30]. Their results showed that pectocellulosic bioplastics showed attractive closed-loop recycling properties. In this context, produced bioplastics from citrus peels using ultrasonic, followed by acetic acid extraction and casting, demonstrated good flexibility and water-stability, and when utilized as packaging, significantly extended the storage time of perishable fruit [18]. Besides citrus peels, the bioplastic synthesis from potato starch and banana peels using acid hydrolysis (HCl) is an environmentally benign and economically effective technology that can be rapidly scaled up for bioplastic manufacturing [21]. Banana peel waste is used for the manufacture of bioplastic utilizing alkaline and acid treatment, or it is further modified to improve its mechanical and physical properties and improve its weight-carrying capabilities [20].

Combining solid-state enzymatic digestion and fermentation resulted in 16% sugar release from grape pomace and a 54% increase in PHA synthesis after 48 hours of fermentation [31]. According to Penkhrue et al. [32], using pineapple peel as a substrate in batch fermentation can yield 5.6 g/L PHB.

3.2 Agricultural residues

Agricultural residues include both field residues (such as stems, leaves, husks, and stalks) and process residues (such as seed leftovers from cotton linters) that leftovers after the crop is converted to its final form [33]. There are some factors affecting bioplastic development by adding natural polymers like lignocellulosic fibers. These factors include formulation and modification steps, as the extraction and purification processes affect the final characteristics of bioplastic and material's economic viability and cost [34]. According to Karaca et al. [35], cellulose fiber and crystal biogenic silica from rice husks and produced silica-reinforced cellulose-based composite bioplastic was manufactured for use in packaging, and the produced films exhibited a considerably greater level of biodegradability compared to traditional plastics, with values ranging from 25.5% to 55.7%.

Due to the numerous varied compositions and features of biomass, the selection of biomass for starch synthesis and additional alteration is critical. Cereals such as wheat, corn, and rice are the primary sources of commercial starch, containing over 60% starch content, and the roots or tuberous of cassava and potato contribute about 16–24% starch [34]. Corn starch, a widespread and abundant source of starch, has been employed as a primary gradient for bioplastic manufacture in several prior research [36, 37]. Faruqy and Chang [36] utilized starch from corn, cassava, and potato to create bioplastic sheets that were mixed with recycled newspaper pulp fiber at varying ratios. Their results concluded that the optimum composition of fibers/bioplastic was B50%: N50%, resulting in the highest percentage of tensile strength. In the study conducted by Shafqat et al. [37], the production of thermoplastic corn starch films in collaboration with various microalgae species (*Nannochloropsis*, *Spirulina*, and *Scenedesmus*) as fillers can improve the physical and chemical features of the produced bioplastic films.

One of the most effective methods for the production of bioplastics is the extraction or conversion of renewable resources into nutritional supplements for microbial fermentation, and this process produces the desired products, including LA and PHA [19]. Pretreatment of the substrate, fermentation accumulation, and extraction are the main phases in microbial fermentation [19]. PLA is typically created by polycondensation of LA or ring-opening polymerization of lactide, while sugar fermentation is the only mechanism that creates LA [19]. In LA fermentation investigated by Morão and de Bie [38], the pH of the culture broth is regulated to induce the formation of calcium lactate, and then calcium lactate is acidified using H_2SO_4, facilitating the extraction of crude LA. PLA was produced through the polymerization of lactide, which is produced from purified LA, and the remaining lactide monomer is removed and recycled within the PLA production process. PLA causes less environmental stress than petroleum-based plastics [38].

3.3 Algae-based sources

According to de Castro et al. [39], microalgae show great promise as bioplastic producers because of their rapid growth and tolerance of many environments. Algae have the potential to overcome the challenges associated with agricultural-based bioplastic production by increasing the global photosynthetic capacity, reversing desertification, and converting carbon dioxide into raw materials [40]. The concept of a third-generation biorefinery, which utilizes microalgae biomass instead of conventional terrestrial plant-based lignocellulosic feedstock, has attracted significant interest [17]. Microalgae have basic characteristics such as high biomass content, fast growth, no need for arable land, and minimal nutritional needs [41]. *Chlorella* has 51–58% protein and *Spirulina* 46–63%, making them suitable microalgae for bioplastic production [17]. Both *Chlorella* and *Spirulina* have small cell sizes and high protein compositions, which make them ideal for film and fiber production work, where particle size is a limiting factor [17]. Furthermore, it is suited for bioplastic conversion without any pretreatment phase, resulting in cost-effective, large-scale production while lowering waste generation [42]. Thermomechanical polymerization of *Chlorella* and *Spirulina* protein biomass to synthesize algal-based bioplastics and thermoplastic blends was investigated by Zeller et al. [42]. Their studies showed that *Chlorella* had stronger bioplastic characteristics than *Spirulina*, although it was less compatible with polyethylene compared with *Spirulina*.

Renewable sugars were obtained from the residue of degreased microalgae biomass (DAB) to produce PHA [26]. The underutilized DAB samples undergo hybrid pretreatment, which involves an acid-catalyzed physicochemical treatment followed by enzymatic pretreatment to achieve the highest possible sugar solubility (0.590 g sugars/g DAB). The combination of conventional non-biodegradable plastics and microalgae produces hybrid composite materials that have the potential to facilitate the recycling of unused conventional plastics [17].

4. Environmental impact and benefits of bioplastics

Regarding packaging polymers, biodegradable technologies represent an excellent and promising approach to meet environmental constraints without disadvantages to other known features of polymers. Concern about environmental pollution caused by the extensive use of synthetic and non-biodegradable materials has prompted researchers to exploit natural resources in a bid to develop alternative eco-friendly polymers. Biodegradable polymers that are obtained from renewable resources represent a viable alternative to synthetic polymers and could augment or replace them to solve natural resources safeguard and environmental pollution. Biodegradable materials are not a solution for all waste problems, but they could be an important aspect of a waste minimization plan. Biodegradability, if suitably controlled, can achieve considerable environmental advantages over the recovery and recycling of waste, and their total lack of recycling would not pose irresolvable waste problems [7].

Material recycling is not a solution to environmental problems, because, as a result, the recycling of materials is frequently not feasible. The food packaging is often irreversibly corrupted; particularly as a result of fatigue, the resulting packaging consists of a material of a low molecular mass and/or unstable compositions that are hard to reclaim to achieve the required quality [3]. Over the last few years, food packaging waste that is generated as a result of population growth has reached worrying quantities, and its degradation is a serious environmental problem. Thus, the development and utilization of packaging materials coming from renewable resources is a source of great interest. The developments of regulations and the ecological sensitivity of consumers have turned our interest toward packaging that is based on new materials. Recycling of materials offers a significant reduction of energy and raw material use, but it is always host to problems [1].

According to Matthews et al. [43], the amount of plastic produced globally increased by 64% from 1975 to 2019, reaching over 338 million tons. Most plastics are not biodegradable, and full decomposition can take centuries [44]. Producing bioplastics from biomass resources like corn, starch, sugarcane, and lignocellulosic substances promotes a circular economy, reduces the need for extracting fossil resources, has a lower carbon footprint, and has the chance to alleviate environmental burdens that occur at the end-of-life [45]. The environmental sustainability of bioplastics, as opposed to synthetic plastics, can be understood in terms of some factors including bioplastic degradability, compostability, and its less impact on the environment concerning climatic change and the dependence on fossil fuels.

4.1 Biodegradability of bioplastic

Biodegradability is a term used to describe the ability of bioplastics (**Figure 2**) to decompose when exposed to certain microorganisms found in nature, including bacteria, algae, and fungi [46]. The process involves two steps: the first fragmentation of long polymer chains into oligomers and monomers, followed by their mineralization by microorganisms resulting in the production of methane, carbon dioxide, water, and biomass [47]. This process can occur in various environments, such as soil, water, or physiological circumstances, and is influenced by several parameters, including the chemical composition of the polymers, the presence of oxygen and light, pH levels, temperature, humidity, microorganisms, and enzymes [46].

4.1.1 Biodegradability of bioplastic in the soil

Biodegradable plastics are those that can biodegrade faster than conventional plastics, to reduce plastics' negative environmental effects and stability [48].

Biodegradable plastics are those that have the potential to biodegrade faster than conventional plastics, intending to reduce the negative environmental effects and environmental persistence of plastics [48]. The bioplastic degradation rate in soil strata is influenced by the type and composition of the bioplastic, environmental conditions (pH, temperature, oxygen availability, moisture content, etc.), microbial diversity and activity, and the presence of additives [49]. Biodegradability studies were performed on bioplastic films produced from untanned proteinaceous trimming waste, which were reported to degrade by 62% within 70 days of soil burial [50]. Additionally, a bioplastic made from rice straw using a Naviglio extractor decomposed fully after 105 days in soil [51].

The starch-based bioplastics derived from maize starch exhibited an 81% degradation rate, whereas the composite including titanium dioxide had a 64% degradation

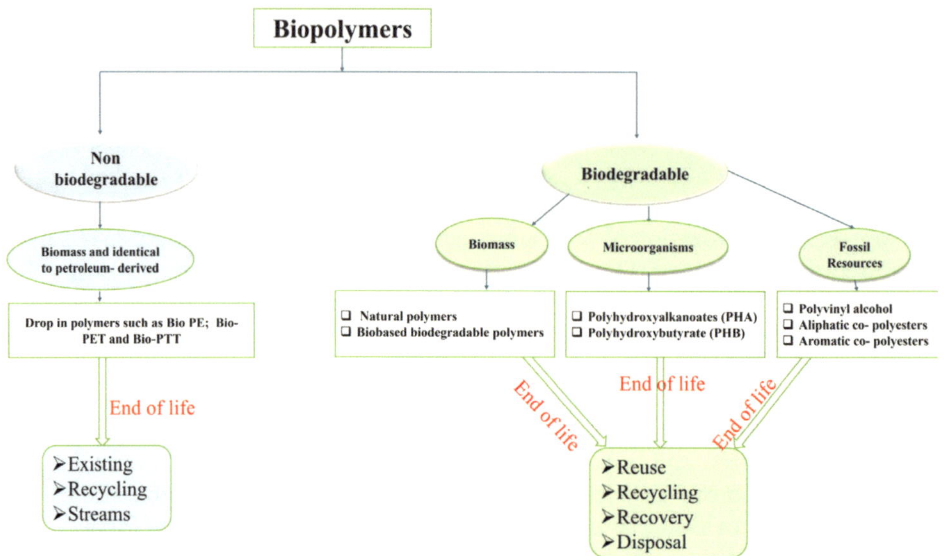

Figure 2.
Biopolymer categorization based on biodegradability and their end of life.

rate during a 1-month soil burial test [22]. Additionally, Starch-based plastics exhibit greater weight reduction and faster degradation in field soil compared to polyhydroxyalkanoates (PHAs) and PLA; however, PLA maintains its weight for an extended duration of approximately 12 weeks [52]. Cellulose-based bioplastics demonstrated the highest level of biodegradability, ranging from 80% to 100%, after 100 days [53]. The study conducted by Ghasemlou et al. [54] explored the biodegradation of non-isocyanate polyhydroxyurethanes (PHUs) in soil and whether, as well as the degradation of PHUs within starch bioplastics (ST) can improve the biodegradation of ST-PHU hybrids. ST-PHU hybrids demonstrated excellent biodegradability, with up to ~88% mass loss after only 120 days of soil burial. Under identical conditions, ST-alone bioplastics degraded at a rate of 69% [54]. A field study on the impact of pure PLA bio-microplastics (bio-MPs) on oat (*Avena sativa* L.) and soybean (*Glycine max* (L.) Merr.) growth and soil health found no significant impact on soil enzyme activities, physicochemical properties, root characteristics, plant biomass, and crop yield [55]. PLA-based bio-MPs may not represent a substantial danger to agroecosystem functions in the near term (days to months) in the field, therefore, they could be a good, eco-friendly alternative to the old, non-biodegradable plastics [55].

4.1.2 Biodegradability of bioplastic in the water

Temperature, salinity, pH, sediments that cause abrasion, bioplastic-degrading microorganisms, and polymer properties affect bioplastic degradation in aquatic environments [49]. Bagheri et al. [56] studied the degradation of several biodegradable plastics, including poly(lactic-co-glycolic acid) (PLGA), PCL, PLA, PHB, Ecoflex, and one synthetic polymer; non-degradable polymer poly(ethylene terephthalate) (PET), in freshwater and artificial seawater in controlled conditions for 1 year. Their results revealed that only PLGA experienced 100% bulk degradation within 270 days because of its amorphous structure, which allowed water molecules to diffuse and accelerate the degradation process.

The microplastic synthesis from biodegradable PBAT was compared with the non-biodegradable Low-Density Polyethylene (LDPE) in Milli-Q water and seawater [57]. They found that under all aquatic circumstances, PBAT produced noticeably more plastic fragments than LDPE. UV-A pretreatment, which simulates the aging of plastic products in sunlight, enhanced the production of PBAT microplastics through oxidation, leading to both crosslinking and chain scission. Furthermore, microplastic production was shown to be quicker in seawater compared with Milli-Q water due to the favorable necessary state for degradation.

Starch blends and Polylactic acid are two of the most prevalent bio-based biodegradable plastics in the worldwide bioplastic industry, contributing 17.9% and 20.7%, respectively [58]. Furthermore, fossil-based LDPE plastic bags with degradant chemicals are widely available on the market in many countries [48]. Cheung and Not [48] used three categories of biodegradable plastics: additive-modified-LDPE, PLA, and polyvinyl alcohol (PVA)/starch mixtures. According to their results, the majority of them fail to degrade completely but remain visible even after a period of 6 months; only PLA exhibits complete degradation in 1–3 months in marine environments, indicating that PLA degradation may be more favorable in subtropical marine climates. Compared to fossil-based plastics (modified-LDPE), bio-based polymers (PLA and PVA/Starch blends) exhibit greater mass losses, ranging from 23% to 100%. Blends of PLA and PVA/cassava starch have better degradation efficiency in marine environments compared to open-air settings (mass losses greater by 50% and 39–41%, respectively), attributed to biodegradation and hydrolysis. Thus, the efficiency degradation process can be influenced by weathering processes, microorganisms, and polymer characteristics.

4.2 Compostability of bioplastic

Bioplastic is biodegradable but not compostable if its kinetics are too slow to meet the compostability standards (home and industrial) [59]. According to the ASTM's definition of compostable plastics, when these polymers decompose, they produce water, carbon dioxide, inorganic compounds, and biomass without leaving any observable or hazardous residues [53]. The composting process is especially suitable for the disposal of food-contaminated packaging, as recycling facilities are unable to handle food-contaminated plastics. Additionally, the compost that is produced can be utilized to enhance soil quality [53]. Composting at high temperatures (over 55°C) causes bioplastics to reach their glass transition temperature, leading to more hydrophilic amorphous polymers and faster degradation [53]. Bioplastics biodegrade almost fully in industrial composting environments that operate between 60°C and 70°C and 60% moisture content [53].

Certain materials require the elevated temperature of composting for timely degradation, these materials can be categorized as "industrial compostable" but not degradable in the open environment, particularly in tough conditions like marine systems, for example, PLA [59]. PLA decomposition in compost occurs only in the presence of compatible microorganisms and favorable environmental conditions, such as high temperatures and humidity. The degradation of PLA in compost is dependent upon the existence of appropriate microorganisms and favorable environmental conditions, such as high temperatures and humidity [46]. The temperature is the primary limiting factor, as the increased flexibility of the chains necessary for chemical and biological degradation only occurs beyond the glass transition temperature of PLA (55°C) [60], whereas, the PLA degradation degree reached 90% within 4 months [46].

Starch is reported in the literature to have good compostability. Gioia et al. [46], showed that after 44 days of incubation under aerobic conditions, 100% of the corn starch at 58°C had mineralized. German et al. [61] reported on the effects of starch concentration on the degradation process based on both laboratory and *in situ* experiments. They found that the mineralization rate of starch, a plant polymer that is frequently present in litter and soil, is concentration-dependent, specifically, its decomposition rate may be decreased by up to 50% when it represents less than approximately 10% of SOM. Lavagnolo et al. [62] investigated the degradation of starch-based bioplastics throughout the composting process. The findings indicated that fragmentation primarily happened during the thermophilic phase, on the other hand, biodegradation happened during the curing phase. The composting process of the waste matrix was completed in 55 days; however, the degradation of bioplastics did not meet the regulatory standards for compostability assessment (≤10% sized >2 mm). The experimental findings demonstrated that a composting duration of 100 days is necessary to fulfill the standard criteria for the compostability of bioplastics. The experimental findings demonstrated a direct correlation between the fragmentation process and a time frame of 100 days would be necessary to fulfill the regulatory criteria.

Similar to starch, cellulose is a polysaccharide that is produced from glucose monomers; however, in cellulose structures, these monomers are more resistant to decomposition due to the presence of stronger glycosidic bonds [53]. Regarding PBAT, several studies have specifically examined the bacteria that can degrade PBAT in various composting environments, including soil, home, and industrial composting. They concluded that thermophilic bacteria can efficiently break ester bonds at high temperatures (50–60°C); however, lower temperatures (25–30°C) is slow and is primarily carried out by fungal strains or mesophilic bacteria like firmicutes and proteobacteria [46].

4.3 The effect of bioplastic on reducing carbon footprint

Besides bioplastic degradability and compostability, it can also help lower net CO_2 emissions. Plants cultivated for producing bioplastics can absorb the same amount of CO_2 as they emit during biodegradation [63, 64]. Bioplastics derived from second-generation biomass, such as forest residues, have the potential to substitute petroplastics and mitigate climate change by reducing carbon footprint [65].

Moreover, the utilization of corn-based PLA bioplastics, as opposed to conventional plastic, results in a 25% decrease in greenhouse gas (GHG) emissions [66]. PLA's degradability reduces environmental stress compared to petroleum-based plastics. To further reduce its impact, reduce NO_x- and SO_x-containing fertilizer consumption, expand waste stream, and reduce chemical use, potentially reducing CO_2 footprint from 501 to 909 kg CO_2-eq/ton PLA [67]. In this regard, the life cycle assessment (LCA) findings indicate that the PHA produced from municipal food waste and wastewater sewage sludge has a significantly lower environmental impact, with a reduction of four times (−16 kg CO_2-eq) compared to synthetic LDPE (4 kg CO_2-eq).

Several studies have demonstrated that the utilization of innovative bioplastics derived from renewable energy sources, such as plant biomass, can effectively reduce GHG emissions [68, 69]. In this regard, the carbon, and environmental footprints of conventional polypropylene plastic and polypropylene-based composites: jute fibers, cotton fibers, kenaf fibers, and glass fibers were examined [68]. Their general findings demonstrated that bioplastic fibers had a reduced carbon footprint and a diminished environmental impact compared to polypropylene. More specifically, the addition of cotton, jute, and kenaf fibers at 30% to a polypropylene matrix resulted in a reduction of its carbon footprint by 3%, 18%, and 18%, respectively compared to the conventional polypropylene. Concerning the ecological footprint, there was a decrease of 8.2% and 9.4% for jute and kenaf fibers, respectively, whereas the ecological footprint of cotton fibers increased by 52% [68].

According to Broeren et al. [70], incorporating starch into biodegradable bioplastics production resulted in a reduction of GHG emissions by up to 80% and a decrease in non-renewable energy use by up to 60%. Compared with pure starch, blends including starch residues exhibit a decrease in land utilization (by up to 60%), the potential for eutrophication (by up to 40%), greenhouse gas emissions (by up to 10%), and these indicators associated with the biomass feedstock cultivation phase [70].

5. Challenges and limitations

The shift toward bio-based and biodegradable plastics presents significant environmental advantages. However, it is imperative to effectively tackle economic and ecological obstacles to promote their widespread acceptance and ensure their sustainable integration into the global plastics industry. Every year, millions of tons of underutilized plastics end up in the oceans, causing significant harm to the environment. Non-biodegradable plastics that are disposed of in landfills can have uncertain long-term consequences, which can restrict the amount of land that is available. Plastics have been found to release harmful substances into the environment, including carbon dioxide, greenhouse gases, and cancer-causing polycyclic aromatic hydrocarbons and dioxins. This poses significant health and environmental hazards. There is an increasing need to replace petroleum-based plastics with biopolymers that are biodegradable and renewable to address these risks. This shift has the potential to minimize environmental damage and promote sustainability by decreasing dependence on non-renewable resources [71].

Bioplastics, which are made from renewable resources, are gaining increasing attention from the industry due to the environmental threat posed by petro-plastics and the limited reserves of fossil fuels. Because they resemble synthetic polymers (PHBs) and can be landfilled or composted for disposal, PLA, TPS, and PHB can considered as a special interest material [72]. PLA, PHB, and TPS are predicted to make up just 29% of the bioplastics consumed in the future [73].

One strategy to lower the emissions from the production of plastic is to use biogenic resources to partially replace the current feedstock supplies. Sugarcane, maize, and cassava are common biogenic feedstocks used to make plastics. Utilizing biogenic resources in their production could reduce GHG emissions by as much as 225% compared to conventional plastics [74, 75].

5.1 Production costs and scalability

The shift from using fossil resources to utilizing renewable materials in the production of plastics shows potential for substantial ecological advantages, such as a diminished carbon footprint and reduced dependence on finite fossil fuels. Furthermore, the possibility of biodegradability presents a hopeful resolution to the increasing issue of plastic waste buildup on land and in oceans, strengthening the desire of consumers for bio-based and biodegradable plastics in response to the growing awareness of environmental concerns [72, 76, 77].

Although bio-based and biodegradable plastics offer numerous benefits, they currently represent a mere 1% of the total global plastic production, which amounts to over 368 million tons per year. The limited adoption of this is partially attributed to the increased costs and environmental concerns related to resource extraction [76].

To identify the cost contributions of particular inputs and process stages a production chain perspective is employed. The Monte Carlo technique, applied for the first time in analyzing biopolymers costs, it helps incorporate uncertainties into current estimates. This method considers both technological input requirements and fluctuations in input prices as factors of uncertainty [76].

Wellenreuther et al. [76] conducted the initial meta-analysis of the expenses associated with manufacturing bio-based plastic polymer PLA using two different sources of raw materials: corn grain and corn stover. Although large-scale production of PLA using corn grain has been well-established for a long time, the use of corn stover as an alternative feedstock, which has the potential to save land resources, has not yet reached the stage of widespread production. Corn stover-based PLA shows promise in terms of competitive variable costs, it faces challenges due to higher fixed costs compared to corn grain-based PLA. Despite these challenges, the Monte Carlo simulations underscore a consistent cost advantage for corn stover-based PLA, contingent upon technological advancements and scale-up efficiencies; achieving competitiveness in the production of bio-based plastics necessitates convincing end users of the environmental benefits associated with switching to alternative feedstocks. This effort may require feedstock-specific adjustments in labeling and taxation, reflecting a holistic approach to policy-making that considers the entire plastic life cycle. However, it is crucial to recognize that corn stover represents only one of several feasible options among innovative feedstocks. To strike an optimal balance between cost competitiveness and ecological impacts, ongoing and consistent life cycle assessments are essential across various second- and third-generation feedstock technologies. This approach is pivotal for guiding future research and industry decisions toward sustainable plastic production. The depletion of fossil resources and the associated environmental concerns have prompted the emergence of a pioneering bioeconomy and facilitated the transition from fossil-based plastics to bio-based plastics. Bio-based plastics have the potential to annually save between 241 and 316 million metric tons of CO_2 equivalent [78].

5.2 Sustainability strategic for bioplastic

The superiority of bioplastics over petro-plastics lies primarily in their sustainability, which can be assessed through life cycle assessment (LCA), life cycle costing (LCC), and social LCA (S-LCA) [72]. For sustainability to be achieved, it is necessary to actively involve societies and markets in envisioning the future role of bioplastics key considerations include improving biodegradability, establishing cost-effective recycling facilities, and optimizing agricultural application. Investigating alternative bioplastic materials has the potential to improve community lifestyles, while lowering recycling expenses could open up new possibilities in various sectors like agriculture and medicine; Biodegradable polymeric materials are expected to emerge as strong substitutes for petrochemical-based plastics in the near future [71].

5.3 Infrastructure for end-of-life management

Biodegradable plastics offer a promising solution for mitigating environmental impacts at the end of their life cycle. Despite currently representing only 1% of the global plastics market, bioplastics are projected to increase their market share by 35% by 2025. Biodegradable alternatives have the potential to replace up to 90% of conventional plastics, but at present, they can only meet 35% of the demand. Bioplastics distinguish themselves from traditional plastics by the significant impact that the way they are disposed of at the end of their life has on their overall emissions throughout their entire life cycle. This is due to the emissions associated with their biodegradation during the disposal process. While petroleum-based plastics typically account for less than 10% of their GHG emissions over their entire life cycle, bioplastics that fully biodegrade in landfills can contribute up to 80% of their life-cycle GHGs, including biogenic emissions. Therefore, considering the expected rise in the usage of bioplastics, it is crucial to comprehensively assess their environmental impacts and develop appropriate strategies for handling their disposal [79, 80].

Managing bioplastic waste presents both opportunities and challenges. Traditional approaches like recycling, incineration, and landfilling are commonly used, but bioplastics offer the additional possibility of being treated as organic waste through composting or anaerobic digestion. These methods can help save landfill capacity and allow for the extraction of energy and nutrients, promoting a circular economy. However, existing waste management systems are not adequately prepared to handle bioplastics, and processes like composting and anaerobic digestion may still generate greenhouse gas emissions [63, 81], In 2015, the majority of plastic waste worldwide was disposed of in landfills (58%), followed by incineration (24%) and recycling (18%) [81]. In order to find the optimal way for bioplastic waste management, it is crucial to evaluate the different waste management strategies and the recyclability of bioplastics, take into account the environmental impacts and potential problems associated related to improper management of the different wastes [63, 81].

It is essential to comprehend the characteristics of biodegradation of bioplastics and to assess the assumptions made in determining the environmental impact of biodegradation. Life Cycle Assessments (LCAs) are a methodical strategy employed to examine and measure the ecological consequences linked to every stage of a bioplastic's life cycle [81]. LCA is a technique for evaluating the economic and environmental impacts of bioplastics. This aids in the comparison of various polymer varieties and the evaluation of the advantages of biopolymers. Bioplastics often exhibit reduced effects on climate change; However, they may have elevated effects on eutrophication and toxicity. Analyzing the complete life cycles of bio-based and petrochemical plastics involves considering their lengthy production and recycling processes, which can be intricate [45, 81, 82]. This technique was developed to attain a "circular economy" [45]. The European circular economy plan advises the adoption of sustainable novel materials and alternative feedstocks, provided that proof demonstrates their superiority over petrochemical plastics [45, 81, 82].

The utilization of LCA for the recycling of bioplastic by thermomechanical recycling indicated an average reduction of 100–130% in greenhouse gas emissions compared to all alternative different methods of bioplastic end-of-life assessment. Nevertheless, there is a contention that the conventional advantages of recycling petroleum-based plastics, such as reducing the manufacturing of new materials, are not as significant when it comes to bioplastics.; where thermomechanical recycling also for PLA led to a reduction of 53% and 33% in greenhouse gas emissions compared to anaerobic digestion (AD) and incineration, respectively [81]. Waste transportation is one of the most important factors affecting the recycling process, especially when dealing with large weights [82].

6. Bioplastics in a circular bioeconomy

The circular economy is founded on the principle of deliberate restoration and regeneration. The objective is to revamp waste disposal systems by minimizing adverse effects through the reimagining of products and services using system-wide innovation.

6.1 Economic and social impacts of bioplastics

The circular economy harnesses renewable energy sources to generate benefits in environmental, economic, and social domains. Hence, the incorporation of biological resources, processes, and methods into advanced knowledge-based manufacturing can serve as a highly efficient approach to mitigating plastic waste. This strategy prioritizes the substitution of fossil fuels and the mitigation of GHG emissions. In addition, the production of bioplastics should prioritize generating benefits for regional and local stakeholders, such as governments, investors, employees, and consumers, and this is regarded as an essential component of the bioplastic procedure process [45].

There are different options for the bioplastic end-of-life, such as mechanical and chemical recycling, also solvent-based recycling, industrial composting, direct fuel substitute in plants, burning with and without energy recovery, digestion in an anaerobic, landfill with and without recovery by energy [83].

6.2 Closed-loop systems and resource efficiency

Bioplastics can be recycled in four ways: (1) by reusing the polymer components in a closed-loop system; (2) by mechanically recycling them into simpler products; (3) by breaking them down chemically into monomeric units; and (4) by recovering energy by burning. Every one of these approaches has unique benefits and drawbacks. Closed-loop recycling, for instance, can usually handle garbage that is primarily impure-free [65, 84].

Since there are several methods for recycling the different types of bio-based plastics, the technologies used should be appropriate for the properties of the biopolymers. For a given bioplastic, the best recycling strategy should be to reuse it wherever possible, then recycle it mechanically till its physical characteristics degrade, and ultimately undergo chemical recycling to retrieve the initial monomers or enhance them into higher-value products [72, 85].

Chemical recycling processes that recover the original monomers or produce more valuable chemical building blocks, such as hydrolysis, aminolysis, glycolysis, and alcoholysis, offer promising prospects. Unfortunately, these procedures are currently too expensive and labor-intensive to be used widely in industry. It is imperative to create an efficient catalyst that can depolymerize plastics into their monomers while protecting critical functional groups. For a thorough investigation of the technologies currently in use for commercial plastic upcycling and chemical recycling [72, 85].

Chemical recycling is widely regarded as the most crucial solution because it has the ability to break down polymer waste into its constituent monomers, which can then be reassembled into polymer materials [86]; while mechanical recycling techniques such as extrusion and injection molding are commonly used for large-scale plastic recycling due to their low processing costs. But this method frequently causes the polymer chain length to shorten, which could lower the plastic's quality. Furthermore, it does not deal with the expanding problem of microplastic waste. Because of these issues, recycled plastics are of lower quality and might not be appropriate for use in applications such as food packaging [72, 85]. Utilizing closed-loop chemical recycling offers a practical resolution to the issue of disposing of synthetic polymers. The process entails converting polymers that are linked together into individual monomers at room temperature, providing significant opportunities for complete retrieval. Furthermore, this procedure has the potential to enhance the mechanical characteristics of the regenerated polymers; thus, closed-loop chemical recycling plays a crucial role in the circular polymer economy. Nevertheless, the challenge of creating dynamic bonds that can form strong connections and be easily broken to facilitate the chemical recycling of cross-linked polymers is still a major obstacle [87, 88].

Recent advancements have significantly improved the development of chemically recyclable polymers by designing monomers to favor depolymerization over polymerization. This progress aims to selectively break down polymers into monomers and establish a closed-loop for chemical recycling [86].

There are two ways that can be used to categorize the closed-loop recycling process for innovative thermosets and their downstream components. The initial approach entails hydrolyzing the thermosets by employing an acidic, alkaline, or catalytic solution. This enables the retrieval of monomers or oligomers, which can subsequently be recycled to manufacture thermosets or their functional materials. The second approach is disrupting the intermolecular connections in the thermosets and subsequently reconstructing them via the exchange reaction of dynamic bonds. Recently, several types of bonds that can be easily broken, including ester bonds, acetal linkages, bases of Schiff, bonds of disulfide, boronic ester links, and hexahydrotriazine structures, have been developed to produce thermosets that can be efficiently recycled inside a closed-loop system.

Through the use of closed-loop recycling via hydrolysis, the thermosets that are recovered can achieve a structure and properties that are exactly the same as the original resin. Nevertheless, the degree to which the degradation process is successful relies significantly on factors such as the ability of the resin to be wetted and the pH level of the solution or catalyst. As a result, attaining close to 100% recovery of the monomer is a difficult undertaking. On the other hand, the closed-loop recycling of thermosets through dynamic reactions aims to achieve a recovery rate of 100%. However, this approach necessitates operating at higher temperatures, and the recycled resin generally exhibits inferior performance compared to the original thermosets [89]. Liu et al. [89] found that the economic benefits and decrease in CO_2 emissions resulting from closed-loop recycling via hydrolysis will serve as major factors in promoting the recycling of thermoset waste in the future. To enhance cost-effectiveness, it is important for the future of closed-loop recycling through dynamic exchange reactions to focus on improving the efficiency of depolymerization and developing milder depolymerization conditions.

Watanabe et al. [90] showed that closed-loop recycling using Aqueous emulsion polymerization resulted in poly methyl acrylate (poly-MA) microparticle films with exceptional resistance to pulling and piercing forces, and the tensile test-based quantitative stress-strain analysis showed that the poly-MA-microparticle films had fracture energy (a measure of a material's toughness) that was on par with or higher than that of traditional poly-MA bulk films. These films also have a fracture energy that is higher than that of other resilient latex films with chemically cross-linked particle interfaces, as well as natural rubber-latex films.

7. Conclusion

Biodegradable polymers offer a promising avenue to reconcile human needs and environmental concerns. They possess the potential to address the escalating ecological challenges created by non-biodegradable polymers while maintaining the convenience and functionality that society demands. These materials, derived from sustainable sources, help decrease plastic pollution and promote resource efficiency through closed-loop systems. Implementing strategic sustainability strategies, such as improving the ability of materials to break down naturally and optimizing the way products are made, is crucial for overcoming these limits. By incorporating biodegradable polymers into circular bioeconomy frameworks, we can enhance the sustainability of material cycles and minimize environmental harm. It is essential to adopt these advancements in order to attain a future that is more robust and environmentally pleasant, with minimal waste and efficient resource management. As we continue to explore and innovate, it is crucial to prioritize research, development, and widespread adoption of biodegradable materials as a sustainable alternative to preserve our planet for future generations.

Conflict of interest

The authors declare no conflict of interest.

Author details

Heba Younis, Fatma Abdelrahman, Mohamed Samer and Hassan Abdellatif*
Agricultural Engineering Department, Faculty of Agriculture, Cairo University, Giza, Egypt

*Address all correspondence to: hassan_elshemy2009@cu.edu.eg

References

[1] Vandermeulen GWM, Boarino A, Klok H-A. Biodegradation of water-soluble and water-dispersible polymers for agricultural, consumer, and industrial applications—Challenges and opportunities for sustainable materials solutions. Journal of Polymer Science. 2022;**60**(12):1797-1813

[2] Gamage A, Liyanapathiranage A, Manamperi A, Gunathilake C, Mani S, Merah O, et al. Applications of starch biopolymers for a sustainable modern agriculture. Sustainability. 2022;**14**(10):6085. DOI: 10.3390/su14106085

[3] Berninger T, Dietz N, González LÓ. Water-soluble polymers in agriculture: Xanthan gum as eco-friendly alternative to synthetics. Microbial Biotechnology. 2021;**14**(5):1881-1896

[4] Rosenboom J-G, Langer R, Traverso G. Bioplastics for a circular economy. Nature Reviews Materials. 2022;**7**(2):117-137

[5] Daria M, Krzysztof L, Jakub M. Characteristics of biodegradable textiles used in environmental engineering: A comprehensive review. Journal of Cleaner Production. 2020;**268**:122129

[6] Alaswad SO, Mahmoud AS, Arunachalam P. Recent advances in biodegradable polymers and their biological applications: A brief review. Polymers (Basel). 2022;**14**(22):4924. DOI: 10.3390/polym14224924. PMID: 36433050; PMCID: PMC9693219

[7] La Fuente CIA, Maniglia BC, Tadini CC. Biodegradable polymers: A review about biodegradation and its implications and applications. Packaging Technology and Science. 2023;**36**(2):81-95

[8] Mo A, Zhang Y, Gao W, Jiang J, He D. Environmental fate and impacts of biodegradable plastics in agricultural soil ecosystems. Applied Soil Ecology. 2023;**181**:104667

[9] Tyagi P, Agate S, Velev OD, Lucia L, Pal L. A critical review of the performance and soil biodegradability profiles of biobased natural and chemically synthesized polymers in industrial applications. Environmental Science and Technology. 2022;**56**(4):2071-2095

[10] Mansoor Z, Tchuenbou-Magaia F, Kowalczuk M, Adamus G, Manning G, Parati M, et al. Polymers use as mulch films in agriculture—A review of history, problems and current trends. Polymers. 2022;**14**(23):5062. DOI: 10.3390/polym14235062

[11] Maraveas C. Production of sustainable and biodegradable polymers from agricultural waste. Polymers. 2020;**12**(5):1127. DOI: 10.3390/polym12051127

[12] Verma K, Sarkar C, Saha S. Biodegradable polymers for agriculture. In: Saha S, Sarkar C, editors. Biodegradable Polymers and Their Emerging Applications. Singapore: Springer Nature Singapore; 2023. pp. 191-212

[13] Akhir MAM, Mustapha M. Formulation of biodegradable plastic mulch film for agriculture crop protection: A review. Polymer Reviews. 2022;**62**(4):890-918

[14] Rai P, Mehrotra S, Priya S, Gnansounou E, Sharma SK. Recent

advances in the sustainable design and applications of biodegradable polymers. Bioresource Technology. 2021;**325**:124739

[15] Volova TG, Uspenskaya MV, Kiselev EG, Sukovatyi AG, Zhila NO, Vasiliev AD, et al. Effect of monomers of 3-hydroxyhexanoate on properties of copolymers poly(3-hydroxybutyrate-co 3-hydroxyhexanoate). Polymers. 2023;**15**(13):2890

[16] Obulisamy PK, Mehariya S. Polyhydroxyalkanoates from extremophiles: A review. Bioresource Technology. 2021;**325**:124653

[17] Roy Chong JW, Tan X, Khoo KS, Ng HS, Jonglertjunya W, Yew GY, et al. Microalgae-based bioplastics: Future solution towards mitigation of plastic wastes. Environmental Research. 2022;**206**:112620

[18] Zhang S, Cheng X, Yang W, Fu Q, Su F, Wu P, et al. Converting fruit peels into biodegradable, recyclable and antimicrobial eco-friendly bioplastics for perishable fruit preservation. Bioresource Technology. 2024;**406**:131074

[19] Li H, Zhou M, Mohammed AEGAY, Chen L, Zhou C. From fruit and vegetable waste to degradable bioplastic films and advanced materials: A review. Sustainable Chemistry and Pharmacy. 2022;**30**:100859

[20] Choudhary P, Pathak A, Kumar P, S C, Sharma N. Commercial production of bioplastic from organic waste–derived biopolymers viz-a-viz waste treatment: A minireview. Biomass Conversion and Biorefinery. 2024;**14**(10):10817-10827

[21] Rizwana Beevi K, Sameera Fathima AR, Thahira Fathima AI, Thameemunisa N, Noorjahan CM, Deepika T. Bioplastic synthesis using banana peels and potato starch and characterization. International Journal of Scientific and Technology Research. 2020;**9**(1):1809-1814

[22] Amin MR, Chowdhury MA, Kowser MA. Characterization and performance analysis of composite bioplastics synthesized using titanium dioxide nanoparticles with corn starch. Heliyon. 2019;**5**(8):e02009

[23] Mostafa NA, Farag AA, Abo-dief HM, Tayeb AM. Production of biodegradable plastic from agricultural wastes. Arabian Journal of Chemistry. 2018;**11**(4):546-553

[24] Li J, Yang Z, Zhang K, Liu M, Liu D, Yan X, et al. Valorizing waste liquor from dilute acid pretreatment of lignocellulosic biomass by bacillus megaterium B-10. Industrial Crops and Products. 2021;**161**(932):113160

[25] Yaradoddi JS, Banapurmath NR, Ganachari SV, Soudagar MEM, Sajjan AM, Kamat S, et al. Bio-based material from fruit waste of orange peel for industrial applications. Journal of Materials Research and Technology. 2021;**17**:3186-3197

[26] Naresh Kumar A, Chatterjee S, Hemalatha M, Althuri A, Min B, Kim SH, et al. Deoiled algal biomass derived renewable sugars for bioethanol and biopolymer production in biorefinery framework. Bioresource Technology. 2020;**296**:122315

[27] Zoungranan Y, Lynda E, Dobi-Brice KK, Tchirioua E, Bakary C, Yannick DD. Influence of natural factors on the biodegradation of simple and composite bioplastics based on cassava starch and corn starch. Journal of Environmental Chemical Engineering. 2020;**8**(5):104396

[28] Asunis F, Carucci A, De Gioannis G, Farru G, Muntoni A,

Polettini A, et al. Combined biohydrogen and polyhydroxyalkanoates production from sheep cheese whey by a mixed microbial culture. Journal of Environmental Management. 2022;**322**:116149

[29] De Laurentiis V, Corrado S, Sala S. Quantifying household waste of fresh fruit and vegetables in the EU. Waste Management. 2018;77:238-251

[30] Zhang S, Fu Q, Li H, Wu P, Waterhouse GIN, Li Y, et al. A pectocellulosic bioplastic from fruit processing waste: Robust, biodegradable, and recyclable. Chemical Engineering Journal. 2023;**463**:142452

[31] Martínez-Avila O, Llenas L, Ponsá S. Sustainable polyhyd-roxyalkanoates production via solid-state fermentation: Influence of the operational parameters and scaling up of the process. Food and Bioproducts Processing. Mar 2022;**132**:13-22. DOI: 10.1016/j.fbp.2021.12.002

[32] Penkhrue W, Jendrossek D, Khanongnuch C, Pathomareeid W, Aizawa T, Behrens RL, et al. Response surface method for polyhydroxybutyrate (PHB) bioplastic accumulation in bacillus drentensis BP17 using pineapple peel. PLoS ONE. 2020;**15**(3):1-21

[33] Gupta N, Mahur BK, Izrayeel AMD, Ahuja A, Rastogi VK. Biomass conversion of agricultural waste residues for different applications: A comprehensive review. Environmental Science and Pollution Research. 2022;**29**(49):73622-73647

[34] Abe MM, Martins JR, Sanvezzo PB, Macedo JV, Branciforti MC, Halley P, et al. Advantages and disadvantages of bioplastics production from starch and lignocellulosic components. Polymers. 2021;**13**(15):2484. DOI: 10.3390/polym13152484

[35] Karaca AE, Özel C, Özarslan AC, Yücel S. The simultaneous extraction of cellulose fiber and crystal biogenic silica from the same rice husk and evaluation in cellulose-based composite bioplastic films. Polymer Composites. 2022;**43**(10):6838-6853

[36] Faruqy R, Chang K. Properties of bioplastic sheets made from different types of starch incorporated with recycled newspaper pulp. Transactions on Science and Technology. 2016;**3**(2):257-264

[37] Shafqat A, Tahir A, Mahmood A, Tabinda AB, Yasar A, Pugazhendhi A. A review on environmental significance carbon foot prints of starch based bio-plastic: A substitute of conventional plastics. Biocatalysis and Agricultural Biotechnology. 2020;**27**:101540

[38] Morão A, de Bie F. Life cycle impact assessment of polylactic acid (PLA) produced from sugarcane in Thailand. Journal of Polymers and the Environment. 2019;**27**(11):2523-2539

[39] de Castro TR, de Macedo DC, de Genaro Chiroli DM, da Silva RC, Tebcherani SM. The potential of cleaner fermentation processes for bioplastic production: A narrative review of polyhydroxyalkanoates (PHA) and polylactic acid (PLA). Journal of Polymers and the Environment. 2022;**30**(3):810-832

[40] Karan H, Funk C, Grabert M, Oey M, Hankamer B. Green Bioplastics as part of a circular bioeconomy. Trends in Plant Science. 2019;**24**(3):237-249

[41] Benedetti M, Vecchi V, Barera S, Dall'Osto L. Biomass from microalgae: The potential of domestication towards sustainable biofactories. Microbial Cell Factories. 2018;**17**(1):1-18

[42] Zeller MA, Hunt R, Jones A, Sharma S. Bioplastics and their thermoplastic blends from Spirulina and Chlorella microalgae. Journal of Applied Polymer Science. 2013;**130**(5):3263-3275

[43] Matthews C, Moran F, Jaiswal AK. A review on European Union's strategy for plastics in a circular economy and its impact on food safety. Journal of Cleaner Production. 2021;**283**:125263

[44] Ali SS, Elsamahy T, Koutra E, Kornaros M, El-Sheekh M, Abdelkarim EA, et al. Degradation of conventional plastic wastes in the environment: A review on current status of knowledge and future perspectives of disposal. The Science of the Total Environment. 2021;**771**:144719

[45] Ali SS, Abdelkarim EA, Elsamahy T, Al-Tohamy R, Li F, Kornaros M, et al. Bioplastic production in terms of life cycle assessment: A state-of-the-art review. Environmental Science and Ecotechnology. 2023;**15**:100254

[46] Gioia C, Giacobazzi G, Vannini M, Totaro G, Sisti L, Colonna M, et al. End of life of biodegradable plastics: Composting versus re/upcycling. ChemSusChem. 2021;**14**(19):4167-4175

[47] Agarwal S. Biodegradable polymers: Present opportunities and challenges in providing a microplastic-free environment. Macromolecular Chemistry and Physics. 2020;**221**(6):2000017. DOI: 10.1002/macp.202000017

[48] Cheung CKH, Not C. Degradation efficiency of biodegradable plastics in subtropical open-air and marine environments: Implications for plastic pollution. Science of the Total Environment. 2024;**938**:173397

[49] Thomas AP, Kasa VP, Dubey BK, Sen R, Sarmah AK. Synthesis and commercialization of bioplastics: Organic waste as a sustainable feedstock. Science of the Total Environment. 15 Dec 2023;**904**:167243. DOI: 10.1016/j.scitotenv.2023.167243

[50] Muralidharan V, Arokianathan MS, Balaraman M, Palanivel S. Tannery trimming waste based biodegradable bioplastic: Facile synthesis and characterization of properties. Polymer Testing. 2020;**81**:106250

[51] Bilo F, Pandini S, Sartore L, Depero LE, Gargiulo G, Bonassi A, et al. A sustainable bioplastic obtained from rice straw. Journal of Cleaner Production. 2018;**200**:357-368

[52] Kawashima N, Yagi T, Kojima K. How do bioplastics and fossil-based plastics play in a circular economy? Macromolecular Materials and Engineering. 2019;**304**(9):1-14

[53] Ahsan WA, Hussain A, Lin C, Nguyen MK. Biodegradation of different types of bioplastics through composting—A recent trend in green recycling. Catalysts. 2023;**13**(2):1-14

[54] Ghasemlou M, Daver F, Murdoch BJ, Ball AS, Ivanova EP, Adhikari B. Biodegradation of novel bioplastics made of starch, polyhydroxyurethanes and cellulose nanocrystals in soil environment. Science of the Total Environment. 2022;**815**:152684

[55] Chu J, Zhou J, Wang Y, Jones DL, Ge J, Yang Y, et al. Field application of biodegradable microplastics has no significant effect on plant and soil health in the short term. Environmental Pollution. 2023;**316**:120556

[56] Bagheri AR, Laforsch C, Greiner A, Agarwal S. Fate of so-called biodegradable polymers in seawater and freshwater. Global Challenges. 2017;**1**(4):1-5

[57] Wei X-F, Bohlén M, Lindblad C, Hedenqvist M, Hakonen A. Microplastics generated from a biodegradable plastic in freshwater and seawater. Water Research. 2021;**198**:117123

[58] Bioplastics E. Bioplastics Facts and Figures. European Bioplastics; 2015

[59] Lackner M, Mukherjee A, Koller M. What are "bioplastics"? Defining renewability, biosynthesis, biodegradability, and biocompatibility. Polymers. 2023;**15**(24):1-26

[60] Brdlík P, Borůvka M, Běhálek L, Lenfeld P. Biodegradation of poly(lactic acid) biocomposites under controlled composting conditions and freshwater biotope. Polymers. 2021;**13**(4):1-15

[61] German DP, Chacon SS, Allison SD. Substrate concentration and enzyme allocation can affect rates of microbial decomposition. Ecology. 2011;**92**(7):1471-1480

[62] Lavagnolo MC, Ruggero F, Pivato A, Boaretti C, Chiumenti A. Composting of starch-based bioplastic bags: Small scale test of degradation and size reduction trend. Detritus. 2020;**12**:57-65

[63] Islam M, Xayachak T, Haque N, Lau D, Bhuiyan M, Pramanik BK. Impact of bioplastics on environment from its production to end-of-life. Process Safety and Environmental Protection. 2024;**188**:151-166

[64] Elsawy MA, Kim K-H, Park J-W, Deep A. Hydrolytic degradation of polylactic acid (PLA) and its composites. Renewable and Sustainable Energy Reviews. 2017;**79**:1346-1352

[65] Lamberti FM, Román-Ramírez LA, Wood J. Recycling of bioplastics: Routes and benefits. Journal of Polymers and the Environment. 2020;**28**(10):2551-2571

[66] Porta R. Plastic pollution and the challenge of bioplastics. Journal of Applied Biotechnology and Bioengineering. 2017;**2**(3):15406

[67] Moro G, Bottari F, Van Loon J, Du Bois E, De Wael K, Moretto LM. Disposable electrodes from waste materials and renewable sources for (bio) electroanalytical applications. Biosensors and Bioelectronics. 2019;**146**:111758

[68] Korol J, Hejna A, Burchart-Korol D, Wachowicz J. Comparative analysis of carbon, ecological, and water footprints of polypropylene-based composites filled with cotton. Jute and Kenaf Fibers Materials. 2020;**13**(16):3541. DOI: 10.3390/ma13163541

[69] Abdullayeva M, Yamil GA. Environmental Impact Assessment of Petroleum-Based Bioplastics. In: World Science. Vol. 5. no. 77. Poland: RS Global; 2022. pp. 1-5. DOI: 10.31435/rsglobal_ws/30092022/7867

[70] Broeren MLM, Kuling L, Worrell E, Shen L. Environmental impact assessment of six starch plastics focusing on wastewater-derived starch and additives. Resources, Conservation and Recycling. 2017;**127**:246-255

[71] Thakur S, Chaudhary J, Sharma B, Verma A, Tamulevicius S, Thakur VK. Sustainability of bioplastics: Opportunities and challenges. Current Opinion in Green and Sustainable Chemistry. 2018;**13**:68-75

[72] Zhao X, Wang Y, Chen X, Yu X, Li W, Zhang S, et al. Sustainable bioplastics derived from renewable natural resources for food packaging. Matter. 2023;**6**(1):97-127

[73] Parveen N, Naik SVCS, Vanapalli KR, Sharma HB. Bioplastic packaging in circular economy: A systems-based

policy approach for multi-sectoral challenges. Science of the Total Environment. 2024;**945**:173893

[74] Benavides PT, Lee U, Zarè-Mehrjerdi O. Life cycle greenhouse gas emissions and energy use of polylactic acid, bio-derived polyethylene, and fossil-derived polyethylene. Journal of Cleaner Production. 2020;**277**:124010

[75] de Oliveira CCN, Zotin MZ, Rochedo PRR, Szklo A. Achieving negative emissions in plastics life cycles through the conversion of biomass feedstock. Biofuels, Bioproducts and Biorefining. 2021;**15**(2):430-453

[76] Wellenreuther C, Wolf A, Zander N. Cost competitiveness of sustainable bioplastic feedstocks—A Monte Carlo analysis for polylactic acid. Cleaner Engineering and Technology. 2022;**6**:100411

[77] Filiciotto L, Rothenberg G. Biodegradable plastics: Standards, policies, and impacts. ChemSusChem. 2021;**14**(1):56-72

[78] Spierling S, Knüpffer E, Behnsen H, Mudersbach M, Krieg H, Springer S, et al. Bio-based plastics—A review of environmental, social and economic impact assessments. Journal of Cleaner Production. 2018;**185**:476-491

[79] Kakadellis S, Harris ZM. Don't scrap the waste: The need for broader system boundaries in bioplastic food packaging life-cycle assessment—A critical review. Journal of Cleaner Production. 2020;**274**:122831

[80] Zheng J, Suh S. Strategies to reduce the global carbon footprint of plastics. Nature Climate Change. 2019;**9**(5):374-378

[81] Van Roijen EC, Miller SA. A review of bioplastics at end-of-life: Linking experimental biodegradation studies and life cycle impact assessments. Resources, Conservation and Recycling. 2022;**181**:106236

[82] Lizundia E, Luzi F, Puglia D. Organic waste valorisation towards circular and sustainable biocomposites. Green Chemistry. 2022;**24**(14):5429-5459

[83] Spierling S, Venkatachalam V, Mudersbach M, Becker N, Herrmann C, Endres H-J. End-of-life options for bio-based plastics in a circular economy—Status quo and potential from a life cycle assessment perspective. Resources. 2020;**9**(7):90. DOI: 10.3390/resources9070090

[84] Zhao X, Korey M, Li K, Copenhaver K, Tekinalp H, Celik S, et al. Plastic waste upcycling toward a circular economy. Chemical Engineering Journal. 2022;**428**:131928

[85] Garcia JM, Robertson ML. The future of plastics recycling. Science. 2017;**358**(6365):870-872

[86] Zhao W, He J, Zhang Y. Chemical closed-loop recycling of polymers realized by monomer design. Fundamental Research. 2024. DOI: 10.1016/j.fmre.2024.05.015. Available online 6 June 2024 (In press)

[87] Qin B, Liu S, Huang Z, Zeng L, Xu J-F, Zhang X. Closed-loop chemical recycling of cross-linked polymeric materials based on reversible amidation chemistry. Nature Communications. 2022;**13**(1):7595

[88] Haque FM, Ishibashi JSA, Lidston CAL, Shao H, Bates FS, Chang AB, et al. Defining the macromolecules of

tomorrow through synergistic sustainable polymer research. Chemical Reviews. 2022;**122**(6):6322-6373

[89] Liu Y, Yu Z, Wang B, Li P, Zhu J, Ma S. Closed-loop chemical recycling of thermosetting polymers and their applications: A review. Green Chemistry. 2022;**24**(15):5691-5708

[90] Watanabe T, Minato H, Sasaki Y, Hiroshige S, Suzuki H, Matsuki N, et al. Closed-loop recycling of microparticle-based polymers. Green Chemistry. 2023;**25**(9):3418-3424. DOI: 10.1039/d3gc00090g